NEW JERSEY ASK3 MATH TEST

Thomas P. Walsh, M.A., M.Ed., Ed.D.
Dan M. Nale, M.Ed.

BARRON'S

Acknowledgments

I would like to thank my wife, Elaine, for her understanding, great criticisms, and support. I would like to thank my sons, Matthew and Dan, who helped me with ideas. I would also like to thank everyone at Kean University who have encouraged me through this project. —Tom Walsh

I would like to thank my wife, Teresa, for her understanding, support, and enthusiasm. Thank you to my parents, Joel and Michelle, for providing a foundation for my successful endeavors. I would also like to thank the Somers Point school district for providing me with the opportunity to become a third-grade teacher. —Dan M. Nale

New Jersey Core Curriculum Content Standards (NJCCCS), set out on pages 2 to 12, are © copyright New Jersey Department of Education.
Format of Sample Tests in Chapter 8 © copyright NJDOE.

All inquiries should be addressed to:
Barron's Educational Series, Inc.
250 Wireless Boulevard
Hauppauge, New York 11788
www.barronseduc.com

ISBN-13: 987-0-7641-3923-9
ISBN-10: 0-7641-3923-1

Library of Congress Catalog Card No. 2007035143

Library of Congress Cataloging-in-Publication Data

Walsh, Thomas P.
    New Jersey ASK grade 3 mathematics / Thomas P. Walsh, Dan M. Nale.
        p. cm.
    Includes index.
    ISBN-13: 987-0-7641-3923-9 (alk. paper)
    ISBN-10: 0-7641-3923-1 (alk. paper)
    1. Mathematics—Examinations—Study guides. 2. Mathematics—Study and teaching (Elementary)—New Jersey. I. Nale, Dan M. II. Title.

    QA43.W346 2008
    510.76—dc22

                                                                    2007035143

Printed in the United States of America
9 8 7 6 5 4 3 2 1

CONTENTS

4 PATTERNS AND ALGEBRA (FUNCTIONS) / 99

5 DATA ANALYSIS, PROBABILITY, AND DISCRETE MATHEMATICS / 123

PROBLEM SOLVING AND OTHER MATHEMATICAL PROCESSES / 155

 ANSWERS TO PRACTICE PROBLEMS / 181

SAMPLE TESTS / 223

INDEX / 277

INTRODUCTION

RATIONALE, GENESIS, AND SPECIFICATIONS OF THE ASSESSMENT

The New Jersey Assessment of Skills and Knowledge 3 (ASK3) assesses students' achievement in the knowledge and skills defined by the New Jersey Core Curriculum Content Standards (NJCCCS) in language arts literacy, mathematics, and science. The assessment was field tested in 2003 and then operationally employed in 2004 to meet the requirements of the No Child Left Behind Act (NCLB). It is a replacement for the Elementary School Proficiency Assessment (ESPA) that was previously given to fourth-grade students.

The third grade ASK is currently a 3-day assessment given in the latter part of March. Two days are dedicated to Language Arts Literacy, and one day is dedicated to Mathematics. There is also a makeup day designated for students who were absent during the designated assessment dates. The entire test is broken down into smaller sections, which are individually timed. Students are allowed to take a small stretch break in between every few sections. The mathematics portion of the assessment incorporates two types of questions. The majority of the assessment is multiple choice (choose the correct response), but there are also a few open-ended questions (give a written response to support and/or explain your answer).

OUTLINE OF THE ASSESSMENT

The ASK3 is based on of the NJCCCS. These standards are aligned with the national standards, which are becoming a foundation for the development of state standards around the country. While this text will greatly assist with ASK3, it will also greatly assist with other third-grade assessments given by other states throughout the country. There are 5 main standards in mathematics according to the NJCCCS (see Chapter 8 for two sample ASK3 tests, which incorporate these five standards and mirror what may be on the ASK3 or similar third-grade state assessment). These five standards incorporate all of the material that students are required to know by the end of third grade.

STANDARD 4.1 (NUMBER AND NUMERICAL OPERATIONS)

All students will develop number sense and will perform standard numerical operations and estimations on all types of numbers in a variety of ways.

Strands and Cumulative Progress Indicators

Building upon knowledge and skills gained in preceding grades, by the end of **grade 3**, students will:

A. Number Sense

1. Use real-life experiences, physical materials, and technology to construct meanings for numbers (**unless otherwise noted, all indicators for grade 3 pertain to these sets of numbers as well**).
   - Whole numbers through hundred thousands
   - Commonly used fractions (denominators of 2, 3, 4, 5, 6, 8, 10) as part of a whole, as a subset of a set, and as a location on a number line

2. Demonstrate an understanding of whole number place-value concepts.
3. Identify whether any whole number is odd or even.
4. Explore the extension of the place-value system to decimals through hundredths.
5. Understand the various uses of numbers.
   - Counting, measuring, labeling (e.g., numbers on baseball uniforms)
6. Compare and order numbers.

B. Numerical Operations

1. Develop the meanings of the four basic arithmetic operations by modeling and discussing a large variety of problems.
   - Addition and subtraction: joining, separating, comparing
   - Multiplication: repeated addition, area/array
   - Division: repeated subtraction, sharing
2. Develop proficiency with basic multiplication and division number facts using a variety of fact strategies (such as "skip counting" and "repeated subtraction").
3. Construct, use, and explain procedures for performing whole number calculations with
   - Pencil and paper
   - Mental math
   - Calculator
4. Use efficient and accurate pencil-and-paper procedures for computation with whole numbers.
   - Addition of three-digit numbers
   - Subtraction of three-digit numbers
   - Multiplication of two-digit numbers by one-digit numbers
5. Count and perform simple computations with money.
   - Cents notation (¢)

6. Select pencil and paper, mental math, or a calculator as the appropriate computational method in a given situation depending on the context and numbers.
7. Check the reasonableness of results of computations.

C. Estimation

1. Judge without counting whether a set of objects has less than, more than, or the same number of objects as a reference set.
2. Construct and use a variety of estimation strategies (e.g., rounding and mental math) for estimating both quantities and the result of computations.
3. Recognize when an estimate is appropriate, and understand the usefulness of an estimate as distinct from an exact answer.
4. Use estimation to determine whether the result of a computation (either by calculator or by hand) is reasonable.

STANDARD 4.2 (GEOMETRY AND MEASUREMENT)

All students will develop spatial sense and the ability to use geometric properties, relationships, and measurement to model, describe, and analyze phenomena.

Strands and Cumulative Progress Indicators

Building upon knowledge and skills gained in preceding grades, by the end of **grade 3**, students will:

A. Geometric Properties

1. Identify and describe spatial relationships of two or more objects in space.
   - Direction, orientation, and perspectives (e.g., which object is on your left when you are standing here?)
   - Relative shapes and sizes

2. Use properties of standard three-dimensional and two-dimensional shapes to identify, classify, and describe them.
   - Vertex, edge, face, side, angle
   - Three-dimensional figures—cube, rectangular prism, sphere, cone, cylinder, and pyramid
   - Two-dimensional figures—square, rectangle, circle, triangle, pentagon, hexagon, octagon
3. Identify and describe relationships among two-dimensional shapes.
   - Same size, same shape
   - Lines of symmetry
4. Understand and apply concepts involving lines, angles, and circles.
   - Line, line segment, endpoint
5. Recognize, describe, extend, and create space-filling patterns.

B. Transforming Shapes

1. Describe and use geometric transformations (slide, flip, turn).
2. Investigate the occurrence of geometry in nature and art.

C. Coordinate Geometry

1. Locate and name points in the first quadrant on a coordinate grid.

D. Units of Measurement

1. Understand that everyday objects have a variety of attributes, each of which can be measured in many ways.

2. Select and use appropriate standard units of measure and measurement tools to solve real-life problems.
   - Length—fractions of an inch ($\frac{1}{4}$, $\frac{1}{2}$), mile, decimeter, kilometer
   - Area—square inch, square centimeter
   - Weight—ounce
   - Capacity—fluid ounce, cup, gallon, milliliter
3. Incorporate estimation in measurement activities (e.g., estimate before measuring).

E. Measuring Geometric Objects

1. Determine the area of simple two-dimensional shapes on a square grid.
2. Determine the perimeter of simple shapes by measuring all of the sides.
3. Measure and compare the volume of three-dimensional objects using materials such as rice or cubes.

STANDARD 4.3 (PATTERNS AND ALGEBRA)

All students will represent and analyze relationships among variable quantities and solve problems involving patterns, functions, and algebraic concepts and processes.

Strands and Cumulative Progress Indicators

Building upon knowledge and skills gained in preceding grades, by the end of **grade 3**, students will:

A. Patterns

1. Recognize, describe, extend, and create patterns.
   - Descriptions using words and number sentences/expressions

■ Whole number patterns that grow or shrink as a result of repeatedly adding, subtracting, multiplying by, or dividing by a fixed number (e.g., 5, 8, 11, . . ., or 800, 400, 200, . . .)

B. Functions and Relationships

1. Use concrete and pictorial models to explore the basic concept of a function.
   ■ Input/output tables, T-charts

C. Modeling

1. Recognize and describe change in quantities.
   ■ Graphs representing change over time (e.g., temperature, height)
2. Construct and solve simple open sentences involving addition or subtraction (e.g., $3 + 6 = $ ___, $n = 15 - 3$, $3 + $ ___ $ = 3$, $16 - c = 7$).

D. Procedures

1. Understand and apply the properties of operations and numbers.
   ■ Commutative (e.g., $3 \times 7 = 7 \times 3$)
   ■ Identity element for multiplication is 1 (e.g., $1 \times 8 = 8$)
   ■ Any number multiplied by zero is zero
2. Understand and use the concepts of equals, less than, and greater than to describe relations between numbers.
   ■ Symbols ($=$, $<$, $>$)

STANDARD 4.4 (DATA ANALYSIS, PROBABILITY, AND DISCRETE MATHEMATICS)

All students will develop an understanding of the concepts and techniques of data analysis, probability, and discrete mathematics and will use them to model situations, solve problems, and analyze and draw appropriate inferences from data.

Strands and Cumulative Progress Indicators

Building upon knowledge and skills gained in preceding grades, by the end of **grade 3**, students will:

A. Data Analysis (or Statistics)

1. Collect, generate, organize, and display data in response to questions, claims, or curiosity.
   - Data collected from the classroom environment
2. Read, interpret, construct, analyze, generate questions about, and draw inferences from displays of data.
   - Pictograph, bar graph, table

B. Probability

1. Use everyday events and chance devices, such as dice, coins, and unevenly divided spinners, to explore concepts of probability.
   - Likely, unlikely, certain, impossible
   - More likely, less likely, equally likely
2. Predict probabilities in a variety of situations (e.g., given the number of items of each color in a bag, what is the probability that an item picked will have a particular color).
   - What students think will happen (intuitive)
   - Collect data and use that data to predict the probability (experimental)

C. Discrete Mathematics—Systematic Listing and Counting

1. Represent and classify data according to attributes, such as shape or color, and relationships.
   - Venn diagrams
   - Numerical and alphabetical order
2. Represent all possibilities for a simple counting situation in an organized way and draw conclusions from this representation.
   - Organized lists, charts

D. Discrete Mathematics—Vertex-Edge Graphs and Algorithms

1. Follow, devise, and describe practical sets of directions (e.g., to add two 2-digit numbers).
2. Explore vertex-edge graphs
   - Vertex, edge
   - Path
3. Find the smallest number of colors needed to color a map.

STANDARD 4.5 (MATHEMATICAL PROCESSES)

All students will use mathematical processes of problem solving, communication, connections, reasoning, representations, and technology to solve problems and communicate mathematical ideas.

Strands and Cumulative Progress Indicators

Building upon knowledge and skills gained in preceding grades, by the end of **grade 3**, students will:

A. Problem Solving

1. Learn mathematics through problem solving, inquiry, and discovery.

2. Solve problems that arise in mathematics and in other contexts (cf. workplace readiness standard 8.3).
   - Open-ended problems
   - Nonroutine problems
   - Problems with multiple solutions
   - Problems that can be solved in several ways
3. Select and apply a variety of appropriate problem-solving strategies (e.g., "try a simpler problem" or "make a diagram") to solve problems.
4. Pose problems of various types and levels of difficulty.
5. Monitor their progress and reflect on the process of their problem-solving activity.

B. Communication

1. Use communication to organize and clarify their mathematical thinking.
   - Reading and writing
   - Discussion, listening, and questioning
2. Communicate their mathematical thinking coherently and clearly to peers, teachers, and others, both orally and in writing.
3. Analyze and evaluate the mathematical thinking and strategies of others.
4. Use the language of mathematics to express mathematical ideas precisely.

C. Connections

1. Recognize recurring themes across mathematical domains (e.g., patterns in number, algebra, and geometry).
2. Use connections among mathematical ideas to explain concepts (e.g., two linear equations have a unique solution because the lines they represent intersect at a single point).
3. Recognize that mathematics is used in a variety of contexts outside of mathematics.

4. Apply mathematics in practical situations and in other disciplines.
5. Trace the development of mathematical concepts over time and across cultures (cf. world languages and social studies standards).
6. Understand how mathematical ideas interconnect and build on one another to produce a coherent whole.

D. Reasoning

1. Recognize that mathematical facts, procedures, and claims must be justified.
2. Use reasoning to support their mathematical conclusions and problem solutions.
3. Select and use various types of reasoning and methods of proof.
4. Rely on reasoning, rather than answer keys, teachers, or peers, to check the correctness of their problem solutions.
5. Make and investigate mathematical conjectures.
   - Use counterexamples as a means of disproving conjectures
   - Verify conjectures using informal reasoning or proofs
6. Evaluate examples of mathematical reasoning and determine whether they are valid.

E. Representations

1. Create and use representations to organize, record, and communicate mathematical ideas.
   - Concrete representations (e.g., base-ten blocks or algebra tiles)
   - Pictorial representations (e.g., diagrams, charts, or tables)
   - Symbolic representations (e.g., a formula)
   - Graphical representations (e.g., a line graph)

2. Select, apply, and translate among mathematical representations to solve problems.
3. Use representations to model and interpret physical, social, and mathematical phenomena.

F. Technology

1. Use technology to gather, analyze, and communicate mathematical information.
2. Use computer spreadsheets, software, and graphing utilities to organize and display quantitative information.
3. Use graphing calculators and computer software to investigate properties of functions and their graphs.
4. Use calculators as problem-solving tools (e.g., to explore patterns, to validate solutions).
5. Use computer software to make and verify conjectures about geometric objects.
6. Use computer-based laboratory technology for mathematical applications in the sciences.[1]

[1]NJCCCS © copyright NJDOE.

SCORING

ASK3 scores are reported as scale scores and performance levels in each of the content areas. The score ranges and their associated performance levels follow:

- 100–199, Partially Proficient
- 200–249, Proficient
- 250–300, Advanced Proficient

The scores of students who are included in the Partially Proficient level are considered to be below the state minimum of proficiency, and those students may be in need of instructional support. Proficient and Advanced Proficient definitions by the State of New Jersey follow.

PROFICIENT

The student performing at the Proficient level demonstrates an understanding and knowledge of basic informational concepts. The student is able to recognize the operations needed to solve problems and communicate his/her solution. The student applies mathematical skills and knowledge to theoretical and real-world situations. The student demonstrates a basic knowledge of geometry and measurement, spatial relationships, and patterns as well as a basic understanding of data analysis, probability, and discrete mathematics.

ADVANCED PROFICIENT

The student performing at the Advanced Proficient level demonstrates the qualities outlined for proficient performance. The student provides explanations that are consistently clear and thorough. The student consistently demonstrates the ability to abstract relevant information and uses a variety of strategies and/or methods in solving a problem. The student demonstrates an understanding of the reasonableness of his/her answers.

WHAT TO BRING

Students are not required to bring any materials to the assessment. Teachers are provided with all of the necessary materials to take the assessment. Those materials may include: two sharpened pencils per student, scrap paper, a calculator (for designated sections of the test), and a manipulative kit (which may include geometric figures, rulers, and other similar devices). Some of these items are only allowed to be used on certain sections of the test.

The teacher will give instructions as to when those items may be used. Scrap paper is collected by the teacher. Even though the test directs the students to bring a book to read in case they finish a section of the test early, we **strongly** advise **against** this. Students prepare all year long to take this assessment so they really should dedicate every possible minute to ensuring they do their best on it. Students are allowed to review answers in the current timed section they are working on. They are not permitted to flip back or forward to a different section.

Chapter 2

NUMBER AND NUMERICAL OPERATIONS

Numbers are all around us. We use and see numbers on a daily basis. Numbers can be used to represent given values, and they can be used in mathematical operations as well. In the grocery store alone, there are many different uses of numbers: weights of items, cost of items, aisles in the store, quantity in a package, and many more.

DEFINITIONS

PLACE VALUE, ESTIMATION, AND OTHER NUMBER SENSE/ NUMERICAL OPERATION TERMS

Every number has what is called a **place value**. The place value of a number is determined by its location in relation to the decimal point. In third grade, students should be familiar with place value to the hundred thousands position and likewise the hundredths position.

Estimation is the process by which one can round a number to a different number that is more easily computed using mental math.

WHOLE NUMBER PLACE VALUE

346,891

The number above is read as three hundred forty-six thousand eight hundred ninety-one.

Note that the word "and" was not used anywhere in the extended notation. The word "and" is commonly said out loud when repeating a long number such as this (and is often inserted into the sentence in place of the comma). However, the word "and" refers to a decimal position in a number, not a comma. (We will discuss decimals in numbers shortly).

In the number 346,891, each digit has a specific place value that is determined by its overall position. The place values are listed as follows:

Number	3	4	6	,	8	9	1
Place Value	Hundred Thousands	Ten Thousands	Thousands	comma	Hundreds	Tens	Ones

To the right of the numeral one is a decimal. If there are no digits to the right of the decimal, the decimal is generally not written. It is assumed to be located at the end of the number. When you read place values from the decimal it begins (going to the left) as ones, tens, hundreds, thousands, ten thousands, and finally hundred thousands. When we discuss decimals shortly, the same rule will be used but this time we will be going to the right of the decimal. In the large whole number 346,891, the place value of each digit is listed underneath each digit. The place value is permanent, or fixed. The place value of each digit gives that specific digit a certain value. For example, the 8 is in the hundreds position. Therefore, the 8 is worth eight hundred because there are 8 of them in the hundreds position. To prove this, you could count by hundreds eight times. At the end you would have an answer of 800. Likewise, the 4 in the ten thousands position; this number would be worth 40,000 because four ten-thousands are a total of 40,000.

WHOLE NUMBER PLACE-VALUE EXERCISES

1. In the number 43,091, which digit is located in the hundreds position?

   A. 3

   B. 0

   C. 2

   D. 9

2. In the number 913,478, which digit is located in the ones position?

   A. 9

   B. 3

   C. 8

   D. 1

3. In the number 830,562, which digit is located in the ten thousands position?

   A. 2

   B. 7

   C. 5

   D. 3

4. In the number 981,037, which digit is located in the hundred thousands position?

   A. 9

   B. 7

   C. 0

   D. 3

5. In the number 471, which digit is located in the hundreds position?

A. 1

B. 4

C. 7

D. 6

Answers on page 181

SIMILARITIES WITH DECIMALS

Just as each digit in a whole number has a place value, so do the digits in a number written as a decimal. The position of the digit in relation to the position of the decimal determines the value of that digit. In third grade, students should know decimal place values to the hundredth position. The place values to the right of the decimal are exactly as they are to the left, except for two important differences. There is no "ones" value, and every place value has a "-ths" on the end of it. So to the right of the decimal, the first digit is in the "tenths" place, the next digit to the right is in the "hundredths" place, and so on. If you can master the order of place value for whole numbers to the left of the decimal, these will be even easier.

For example, let's look at the number:

0.45

The number 0.45 would be read as forty-five hundredths. To read a decimal number you would just say the number as you see it (45 = forty-five) and end with whatever the last place value of the number is (furthest to the right). In this example, the 5 is the furthest to the right, and is in the hundredths position. Therefore, the number is read as forty-five hundredths.

Number	0	.	4	5
Place Value	Ones	Decimal ("and")	Tenths	Hundredths

In the number forty-five hundredths (0.45), the zero is not read aloud as it has no value. It is in the ones position, and if you have zero in the ones position, there is no value. As we discussed earlier, you should be able to see the similarities in place value when you compare the right side of the decimal to the left side. The right side begins with tenths and then continues just as it would on the left side, except that every value ends with a "-ths." That's all there is to it.

DECIMAL PLACE-VALUE EXERCISES

1. In the number 0.71, which digit is located in the ones position?

   A. 0

   B. 7

   C. 1

2. In the number 0.43, which digit is located in the hundredths position?

   A. 0

   B. 4

   C. 3

3. In the number 0.97, which digit is located in the tenths position?

   A. 0

   B. 9

   C. 7

4. In the number 0.14, which digit is located in the hundredths position?

    A. 0

    B. 1

    C. 4

5. In the number 3.02, which digit is located in the ones position?

    A. 3

    B. 0

    C. 2

Answers on pages 181 to 182

COMBINATION PLACE VALUES WITH WHOLE NUMBERS AND DECIMALS

It is possible and common to have a number that contains both whole numbers and decimal values. In essence, it would be a combination of both items previously learned. The difficulty is no greater than just combining the two ideas that you have previously learned into one problem. For example:

Number	4	1	7	,	3	9	0	.	6	5
Place Value	Hundred Thousands	Ten Thousands	Thousands	comma	Hundreds	Tens	Ones	Decimal ("and")	Tenths	Hundredths

As you can see, the place values have all remained the same as they were previously. In this example, we have merely combined whole numbers and decimals into one large number. This number is read as four hundred seventeen thousand, three hundred ninety and sixty-five hundredths. Remember that after you read the "and" for the decimal, you should group all of those numbers together and read them as you see them. Then add in the place value of the last digit (in this case the 5 is the last digit). This is why we ended the read-aloud as "hundredths."

COMBINATION PLACE VALUES WITH WHOLE NUMBERS AND DECIMALS EXERCISES

1. In the number 3,742.91, which digit is located in the hundreds position?

   A. 3

   B. 7

   C. 1

   D. 9

2. In the number 409.62, which digit is located in the ones position?

   A. 4

   B. 0

   C. 6

   D. 9

3. In the number 914,379.02, which digit is located in the tenths position?

   A. 9

   B. 7

   C. 0

   D. 2

4. In the number 72,316.40, which digit is located in the hundredths position?

   A. 0

   B. 3

   C. 4

   D. 6

5. In the number 316.52, which digit is located in the tenths position?

   A. 3

   B. 1

   C. 2

   D. 5

Answers on page 182

ESTIMATING

Estimating is the process by which we round a number to make it easier to compute using mental math. It gives us a rough idea of the value. This can be useful in many situations, such as when you are in the supermarket. In this case, you would generally not have a calculator handy to compute the cost of the items you are purchasing. Using mental math and estimation, it is possible to tabulate (as you shop) to determine what your total should roughly be.

The estimating process is quite simple, but it does follow a few basic rules. Estimation is not an exact science. Unless noted by the question, or the answers to the question, different individuals will estimate numbers to different values, as they see fit and easy to compute using mental math.

When estimating a number, if we are told to estimate it to the nearest hundred, we must look at the digit in the hundreds position. Once we have located that, to estimate this number, we must look to the digit on the right. In this case, the digit to the right is in the tens position. If the digit to the right is 5 or greater, we will round up the number in the hundreds position. If the digit is less than 5 we will leave the number in the hundreds position the same. Then in both cases, each other digit to the right becomes a zero.

For example, in the number 871 the 8 is in the hundreds position. If we were asked to round 871 to the nearest hundred, we would locate the 8, then look to the right. The number 7 is greater than 5 so the 8 must be rounded up to 9. All other digits to the right become 0. So 871, rounded to the nearest hundred, will be 900. This makes sense because 871 is closer to 900 than it is to 800. It is 71 away from 800 (871 − 800) yet only 29 away from 900 (900 − 871).

Likewise, if we were given the number 1,735 and asked to round it to the nearest hundred, we would do the same thing. 7 is in the hundreds position, the 3 to the right of it is lower than 5; therefore, the 7 remains a 7 and all other digits to the right become a 0. So 1,735 rounded to the nearest hundred would be 1,700 (1,735 is closer to 1,700 than it is to 1,800).

In a supermarket, estimating comes in very handy. It is also easy to do with money. This is where personal preference comes into play. Some people may round $1.40 to $1.50 because that can be fairly simple to add to other costs using mental math. Other people may find it easier to round $1.40 to $1.00 because it is even easier to compute against other costs using mental math. Either method of estimation would be correct; however, you must realize that the further away from the original number you are, the further away your final estimation is going to be from the true calculated value. If you round $1.40 to $1.50 and can compute that using mental math when adding other costs, you are only $0.10 away from the actual value. So your final outcome will be pretty close to the actual value. If you rounded the $1.40 to $1.00 you are now $0.40 away from the actual number. Not to mention the fact that you are now rounding down. If you are shopping in a store, it may make more sense to round up more often. You wouldn't want to get to the cash register, having rounded each number, only to find out that the actual cost is far off and you don't have enough money to pay the bill!

ESTIMATION EXERCISES

1. What is 587 rounded to the nearest hundred?

    A. 500

    B. 580

    C. 600

    D. 590

2. What is 3,971 rounded to the nearest thousand?

    A. 3,980

    B. 3,900

    C. 4,100

    D. 4,000

3. Which number is rounded to the hundreds position?

    A. 9,640

    B. 32,800

    C. 5,222

    D. 8

4. Which number is rounded to the nearest dollar?

    A. $8.90

    B. $8.00

    C. $9.01

    D. $13.30

5. Which number would be correct if we were rounding $1.66 to the nearest dollar?

    A. $1.50

    B. $2.00

    C. $1.70

    D. $1.00

Answers on pages 182 to 183

FRACTIONS

DEFINITIONS

A **numerator** is the top value in a fraction. It refers to the number of items you are attempting to represent.

A **denominator** is the bottom value in a fraction. It refers to the total number of items it would take to make that particular item whole

For example, if it took 8 pieces of pizza to make a whole pie, the denominator would always be 8. If 2 were eaten and you wanted to represent what fraction of the pizza was eaten, the numerator would be 2, for a final fraction of $\frac{2}{8}$.

Fractions, like decimals, are a way to represent a number that is not a whole number. A fraction generally will represent only a portion, or part, of an entire object. For example:

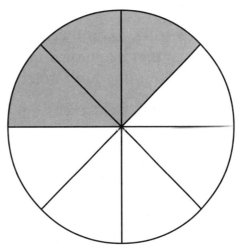

In the pie, you will note that 3 pieces are shaded in. Those shaded pieces represent the amount of pizza that someone ate. You also should realize that the pizza originally was 8 slices in total, before anyone ate any of it.

The pizza, when it arrived, was one whole pizza or $\frac{8}{8}$ (8 slices are there, out of a possibility of 8). Now that someone has eaten 3 slices, there are 5 remaining. So there are two possible fractions with this example.

a. The amount of pizza eaten so far is $\frac{3}{8}$ (or 3 out of a total of 8 original pieces)

b. The amount of pizza remaining is $\frac{5}{8}$ (or 5 out of a total of 8 original pieces)

Similarly, fractions may be used to represent a number line. A number line is a line that is divided into a certain number of sections. Some sections may be labeled with a value, while others are blank, so that you must determine and fill out the appropriate value. For example:

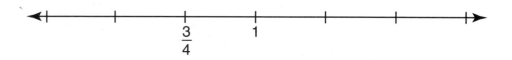

To solve a number line, it is necessary to locate two values that are next to each other. By doing this, we can determine how much each segment of the line is increasing, or decreasing. In this example, $\frac{3}{4}$ and 1 are next to each other. The number 1 is the same thing as $\frac{4}{4}$. Since $\frac{3}{4}$ and $\frac{4}{4}$ are next to each other, we now know that the number line is increasing by $\frac{1}{4}$ in each step going to the right (and oppositely decreasing by $\frac{1}{4}$ in each step going to the left). To determine the first two missing values, all we have to do is start at $\frac{3}{4}$ and subtract $\frac{1}{4}$ each time. The first value to the left of $\frac{3}{4}$ would be $\frac{2}{4}$ (also known as $\frac{1}{2}$), then to the left of that it would be $\frac{1}{4}$. Now we can focus on the two remaining missing values to the right of the 1. The first value to the right would be $1\frac{1}{4}$, the next to the right would be $1\frac{2}{4}$ (or simplified to $1\frac{1}{2}$).

Similarly, if you are given a number line that has no values whatsoever, and you are asked to fill in each of the segments for the value it represents, it will be a similar process. In this example,

to determine the value of each section, you would count up each section (*not* each line) and that would be your denominator. The denominator is the value that represents the total number of segments in a given problem. If the line segment is broken into four segments, or sections, and no values are given, each segment is worth $\frac{1}{4}$. As you count to the right, adding each segment, the values increase.

FRACTION EXERCISES

1. In the diagram,

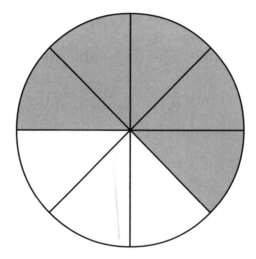

the shaded pieces represent pizza that was eaten. Which value below would represent the fraction of the whole pizza that was eaten?

A. $\dfrac{1}{3}$

B. $\dfrac{3}{8}$

C. $\dfrac{5}{8}$

D. $\dfrac{8}{5}$

2. In the diagram,

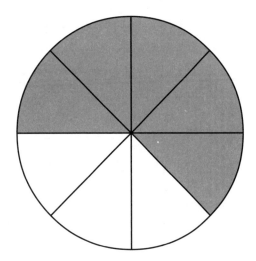

the shaded pieces represent pizza that was eaten. Which value below would represent the fraction of the whole pizza that is uneaten?

A. $\dfrac{5}{8}$

B. $\dfrac{8}{3}$

C. $\dfrac{3}{8}$

D. $\dfrac{1}{8}$

3. In the number line shown,

which fraction would represent the value of each segment?

A. $\dfrac{6}{1}$

B. $\dfrac{1}{1}$

C. $\dfrac{1}{3}$

D. $\dfrac{1}{6}$

4. In the number line shown,

what would be the value of the first two sections combined?

A. $\dfrac{1}{3}$

B. $\dfrac{3}{2}$

C. $\dfrac{2}{3}$

D. $\dfrac{3}{3}$

5. In the diagram shown,

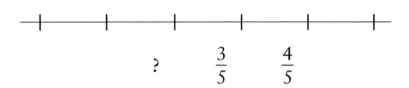

what is the value of the second missing space?

A. $\dfrac{1}{4}$

B. $\dfrac{1}{5}$

C. $\dfrac{2}{5}$

D. $\dfrac{5}{2}$

Answers on pages 183 to 184

IDENTIFYING WHETHER A NUMBER IS EVEN OR ODD

Determining if a large number, or any number for that matter, is even or odd is a very simple process. We all know the single digit even and odd numbers.

For even numbers, we have 0, 2, 4, 6, 8
For odd numbers, we have 1, 3, 5, 7, 9

To determine if a larger number is even or odd, we need to look at the place value furthest to the right. For example, the number 43,921 is odd because the number furthest to the right (1) is odd. Therefore, this is an odd number. The number 813,920 is even because the number furthest to the right (0) is even.

EVEN/ODD EXERCISES

1. Is the number 45,981 even or odd?

    A. Even

    B. Odd

2. Is the number 912,780,312 even or odd?

    A. Even

    B. Odd

3. Which number shown below is even?

    A. 412

    B. 7,231

    C. 89,643

    D. 817

4. Which number shown below is odd?

    A. 500

    B. 5,002

    C. 501

    D. 5,008

5. Which number shown below is even?

    A. 101

    B. 607

    C. 593

    D. 722

Answers on page 184

COMPARING AND ORDERING NUMBERS

Comparing numbers is likewise a very simple task. If you are good at alphabetizing your spelling words, you will find this to be a similarly easy task. When comparing numbers, we must note the place value of each digit in that number. Each place value gives that specific number a final value. For example, in the number 892, the 8 is worth 800, the 9 is worth 90, and the 2 is worth 2. To confirm, if we add those values together, 800 + 90 + 2, we do get the original number of 892. If we wanted to compare this number to 761 we would need to know the value of each digit, and compare them to find out which one is larger. The first thing you should do when comparing numbers is determine if each number has the same amount of digits. If they are different, the larger number will be the one with the greater number of digits. If they have the same number of digits it is a simple process to determine which is greater. In this case, you would do the following:

▪ Look at the digit all the way to the left of each number. In the previous example (892 and 761), we would be looking for the 8 and the 7. Immediately I can determine that 8 is greater than 7; therefore, 892 is greater than 761. Had those two numbers been the same, I would have proceeded to the next digit and compared that.

For example, let's compare the number 801,499 with the number 801,501:

▪ To start off we count the digits in both numbers. We find that they both have six digits, so we will continue to the next step.
▪ Now we will have to look at the digit furthest to the left. Both numbers start with an 8, so we must move to the next digit. Both numbers then have a 0, so we again move to the next digit. Again, both digits then have a 1 so we move over yet another digit. The next digits are a 4 or a 5. 5 is larger than 4, so the larger number is 801,501.

COMPARING NUMBERS EXERCISES

1. Which number is greater than 54,913?
   A. 54,899
   B. 5,979
   C. 54,911
   D. 55,001

2. Which number is greater than 781,401?
   A. 781,701
   B. 781,301
   C. 781,311
   D. 779,876

3. Which statement is true?
   A. 412 is larger than 387.
   B. 5,413 is larger than 6,910.
   C. 154,001 is larger than 499,013.
   D. 10,900 is the same as 1,900.

4. What is the largest even number shown below?
   A. 79,654
   B. 437,711
   C. 79
   D. 694

5. What is the largest odd number shown below?
   A. 101
   B. 608
   C. 593
   D. 722

Answers on page 185

ADDITION

Addition is the arithmetic process of combining or joining numbers together. It is used in many tasks throughout a normal day. Many times, it may not even present itself as an obvious math problem. It may be something you do on a regular basis but never realized you were using math to do it. Mental math may be used to compute many smaller addition problems, but it can also become necessary to use paper and pencil or a calculator to solve the larger ones.

Smaller problems such as 9 + 8 (also known as addition facts because both numbers being added are only one digit) can be done using mental math. These are basic facts that should be memorized to make the larger problems easier to deal with. Using mental math, you should come up with an answer of 17 for 9 + 8.

As the problems grow larger, it will become necessary to use paper and pencil to compute the answer. For example, you will probably need to use paper and pencil to add a two-digit number plus another two-digit number such as 54 + 67. One of the keys to solving addition and even subtraction problems is to always rewrite them vertically. It is of the utmost importance that you *always* rewrite addition problems vertically (one number on top of the other). 54 + 67 is written horizontally (side-by-side) and is difficult to solve because the place values are not lined up on top of each other. Rewriting the problem as

$$\begin{array}{r} 54 \\ + \, 67 \\ \hline \end{array}$$

and lining the place value columns up makes this problem much easier to solve.

In any addition problem, you must always start adding on the right-hand side. So in the example shown above, the first two digits we add are the 4 and the 7. This yields an answer of 11. Underneath the 7, we write the first part of our answer, a 1. The one represents the numeral in the ones position in the number 11. The other 1 (from the tens position) will get carried over on top of the 5 and added to it. Now you are ready to begin adding in the tens position. Here we have the 1 we carried, the 5, and the 6. This gives us a total of 12. Normally we would write the 2 underneath the 6, and carry the 1. However, there is nowhere to carry the 1 to. There are no other digits to the left. If you pretend to carry the 1, you will see that it lands in the hundreds position, which is correct, even though there are no other numerals to add to it. So you can either do this, or just bring it down to the bottom for a final answer of 121.

This process remains the same no matter how many digits you are adding. It could be 94,581 + 8,032. Again, you would rewrite the problem vertically (94,581 on top of the 8,032) and begin adding the digits starting in the ones position and moving over to the left until there is nothing more to add.

$$
\begin{array}{r}
\overset{1}{}\overset{1}{} \\
94{,}581 \\
+\ 8{,}032 \\
\hline
102{,}613
\end{array}
$$

Be absolutely certain that when you rewrite the problem you line everything up on the right-hand side.

ADDITION EXERCISES

1. What is 58 + 9?

   A. 47

   B. 48

   C. 67

   D. 57

2. What is 8,741 + 509?

    A. 9,250

    B. 13,731

    C. 9,520

    D. 9,210

3. Which statement is correct?

    A. 6,413 + 6,413 = 12,816

    B. 5,413 + 801 = 6,214

    C. 154,001 + 4,561 = 159,256

    D. 401 + 339 = 730

4. What is 543 + 399 + 42?

    A. 984

    B. 945

    C. 1,014

    D. 694

5. What is 41 + 32 + 99 + 14?

    A. 192

    B. 206

    C. 184

    D. 186

Answers on pages 185 to 186

SUBTRACTION

Subtraction is the arithmetic process of separating or removing one value from another. Like addition, it is also a process that is commonly used on a daily basis using mental math. Subtraction is only slightly more difficult than addition. The main difference is that borrowing can be involved if the subtraction is not possible (for example, if you are trying to take a larger number from a smaller one, such as 3–7). We will get into that more difficult version shortly.

Subtraction is basically the opposite of addition. If you have the problem $13 + 8 = 21$, you can read it backwards as $21 - 8 = 13$. This problem should be solved using paper and pencil because it requires borrowing. Again, like addition, you should *always* rewrite the problem vertically

$$
\begin{array}{r}
21 \\
-\ 8 \\
\hline
\end{array}
$$

Just like addition, we line the numbers up on the right-hand side, then we begin subtracting on the right-hand side. So we will first compute 1 minus 8. The answer is *not* 7. 1 minus 8 is a negative number. You cannot take 8 from 1. Imagine if you had 1 piece of candy, and a friend wanted to take 8 pieces from you. It isn't possible. So to fix this, we must borrow. Borrowing is the process where you take 1 from the neighboring numeral to the left. In this case, the next numeral is the 2. So cross out the 2, make it a 1 (one less), and add 10 to the original 1 on the right that you were trying to subtract. That makes the new subtraction problem $11 - 8$. This would yield an answer of 3, so write 3 underneath the 8. Now we move to the left. 1 minus nothing is 1. So bring that 1 down. There are no more digits to subtract to the left so our answer of 13 is final.

The borrowing process is by far the most common point of error. You must be very careful to not just borrow the 10 from the neighboring digit, without making it one less. Always show your work each step of the way to avoid silly mistakes.

SUBTRACTION EXERCISES

1. What is 23 – 11?

   A. 11

   B. 12

   C. 2

   D. 34

2. What is 43 – 9?

   A. 43

   B. 34

   C. 52

   D. 46

3. What is 52 – 21?

   A. 21

   B. 73

   C. 31

   D. 33

4. What is 57 – 49?

   A. 12

   B. 106

   C. 41

   D. 8

5. What is 100 – 65?

   A. 45

   B. 25

   C. 35

   D. 40

Answers on pages 187 to 188

MULTIPLICATION

Multiplication is a process very similar to addition. In fact, it is addition on an extended scale. Also like addition, multiplication has a set of basic facts that must be memorized. These basic facts are minimally up to the 9s, and preferably up to the 12s. This means that you should memorize and practice all multiplication facts that are up to 9×9 and/or 12×12. Memorizing these multiplication facts is very important because a good portion of third grade, and basically every grade after that, will rely on these as a foundation for doing other larger math problems. If you don't know the facts, you won't be able to do the future math problems.

MULTIPLICATION FACTS

There are a few tricks to memorizing the facts. Not every numeral has a trick to it; some just require memorization on your part. We'll go through a few of the tricks here:

- For all facts, you can do skip counting. Skip counting is the process by which you count by the given number. For example, if I asked you what 5×3 is, you could count by 5s three times. Five, ten, fifteen; so fifteen is the answer. You could also do it backwards if that is easier for you; count by 3s five times. Three, six, nine, twelve, fifteen; and again we see that 15 is the answer.
- Similar to skip counting is multiplication using arrays. For example, if you were multiplying 4×7 you would make 4 groups of 7 (using dots, or counters, or any other appropriate manipulative) and add them up to get a total of 28.
- Any number times 0 is 0 (for example, $8 \times 0 = 0$).
- Any number times 1 is itself (for example, $9 \times 1 = 9$).
- Any number times 2 is double itself (for example, $6 \times 2 = 12$).
- Any number times 5 will either end in 5 or 0 (for example, $6 \times 5 = 30$, and $5 \times 7 = 35$). How does this

help? Well if you come up with an answer that ends in anything other than a 5 or a 0, you know it is incorrect. If you have a multiple-choice test and some of the choices end in something other than 5 or 0, you can cross them off because they are wrong. This helps narrow down the actual answer.

▪ 6 × 8 is 48 (this one rhymes; six times eight is forty-eight).

▪ There is a neat trick for all 9's. Let's say we're multiplying 9 × 7. Hold both of your hands in the air and spread apart all 10 of your fingers (when we refer to fingers we are including thumbs). This trick *only* works for the 9s facts. Since we're doing 9 × 7, start at your left pinky and count in your head to your seventh finger. You should end at your right index finger (next to your thumb). Put that finger down and keep all the others up. Now, look how many fingers are to the left of the finger you just put down. You should see 6. Now look how many fingers are to the right of the finger you just put down. You should see 3. Put those two numbers (6 and 3) together and you have your answer to 9 × 7; 63. This trick works with any single digit number when multiplied by 9.

▪ There is also a trick you can do by rounding up the number then subtracting. For example, if you did not know that 9 × 7 is 63, you could do 10 × 7, which is a lot easier, and get 70. Then subtract a 7 and get 63. This works because originally we asked how much nine 7's were. If instead you do ten 7's to make it easier, you can just subtract a 7 afterwards and still get the correct answer. Here's another example. 5 × 9 is 45. If you didn't know this immediately, and couldn't figure it out by skip counting, you could instead do 5 × 10 and get 50. Then just subtract a 5 (because you did ten 5s instead of nine 5s) and get 45.

▪ You can make an array with the multiplication fact. If the problem is 3 × 6, you can make an array of dots 3 rows by 6 columns (or vice versa, 6 columns by 3 rows) and count the number of dots.

This way is very tedious and lends itself to mistakes. However, it could be used as a last resort.

Don't forget that multiplication facts can be reversed. 4×7 is the same as 7×4. This is the same as with addition. In subtraction, you *cannot* reverse the numbers and get the same answer.

MULTIPLICATION FACT EXERCISES

1. What is 5×7?
    A. 35
    B. 40
    C. 30
    D. 23

2. What is 9×2?
    A. 16
    B. 18
    C. 13
    D. 9

3. What is 9×7?
    A. 63
    B. 61
    C. 72
    D. 54

4. What is 72×1?

   A. 36

   B. 144

   C. 1

   D. 72

5. What is $7,601 \times 0$?

   A. 7,601

   B. 1

   C. 0

   D. 7,602

Answers on page 188

DIVISION

Division is the process of splitting a number into smaller pieces. For example, if you are given the number 15 and asked to split it into 5 equal pieces, the answer would be 3. You could make 3 equal pieces with 5 in each, totaling 15. Division involves grouping, or equally breaking apart a given number into smaller groups. It can be solved using an elementary method by using counters, or coins. The downside of this method is that it can take a very long time to count out coins, split them into equal groups, and hope that you did not drop one on the floor by accident. Also, you have to be certain you do not give any groups any more than the other groups have (they must be equal). This can also be challenging. Additionally, it can be thought of as the opposite of multiplication. If you are good at multiplication facts, you will probably be good at division facts. To solve smaller division problems just read them backwards as a multiplication problem.

For example, if you were asked the question: $56 \div 8 = ?$ Instead of reading it forward as 56 divided by 8 equals a number, you could read it backwards as "what number times 8 equals 56"? If you know your multiplication facts well, you will realize the answer is 7. The division process here is taking the number 56, splitting it into 8 equal groups and then realizing that in order to do that, you would have exactly 7 in each group with none left over.

Another way to compute a division problem is by skip counting. If you were asked the question: $12 \div 2 = ?$ you could solve this pretty easily with skip counting. Think of it this way, how many times do you have to count by 2s until you reach 12? If you counted aloud, or in your head, you would see that you need to count by 2s six times in order to reach 12. Hence the answer is 6. This method only really works with very small division problems. It becomes difficult to skip count if you are asked a question such as $49 \div 7 = ?$ because it can be confusing to count by 7s. In this case, you would just need to read it backwards as described before and realize that $7 \times 7 = 49$, so the answer is 7. There are seven 7s in 49.

If you were given a slightly more difficult problem such as: $58 \div 8 = ?$ you could similarly solve it backwards. How many groups of 8 can you make with the number 58? If you make 7 groups of 8 you use up 56 of the 58. That leaves a remainder of 2. We can't make another group with that 2 because each group had to have 8 in it. So the final answer becomes 7 remainder 2. So $58 \div 8 = 7$ remainder 2.

In third grade, it is mandated that students learn all of the division facts, similar to that of the multiplication facts. The process of long division begins in fourth grade, where students will hand-write division problems using a longer method (since the division problems will be much larger). So at this time it is important for students to memorize those multiplication facts; this way they will also know their division facts just as well.

DIVISION FACT EXERCISES

1. What is 42 ÷ 6?

   A. 6

   B. 7

   C. 12

   D. 252

2. What is 25 ÷ 5?

   A. 5

   B. 125

   C. 3

   D. 4

3. What is 63 ÷ 7?

   A. 461

   B. 8

   C. 7

   D. 9

4. What is 73 ÷ 8?

   A. 9 remainder 1

   B. 576

   C. 12 remainder 2

   D. 8 remainder 1

5. What is 57 ÷ 9?

   A. 8 remainder 1

   B. 6 remainder 3

   C. 6 remainder 2

   D. 5 remainder 3

Answers on page 189

GEOMETRY AND MEASUREMENT

Geometric shapes are everywhere in our world. We play with lots of different sized balls, which are spheres. Sugar cubes are just that: cubes. We write on pieces of paper, which most of the time are shaped like rectangles. We eat pie, which is shaped like a circle. Along the road, we see traffic signs in many shapes (triangle, square, octagon). In this chapter, we'll explore some characteristics of geometry.

DEFINITIONS

POLYGONS, VERTICES, AND OTHER GEOMETRIC TERMS

A **polygon** is a many sided two-dimensional figure. In other words, it is a closed figure of three or more straight lines on a flat piece of paper. Examples are a triangle, a pentagon, and an octagon.

An **edge** is a side of the polygon. Edges is the plural of "edge." An edge is a line segment, that is, a small piece of a straight line.

A **vertex** is where two edges meet. Vertices is the plural of "vertex." A vertex forms an angle on the polygon.

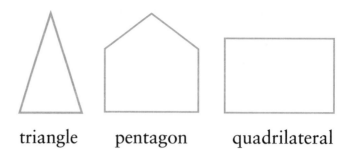

triangle   pentagon   quadrilateral

POLYGONS

There are many polygons. The following table names just a few and tells how many edges and vertices each one has.

Polygon	Edges	Vertices
Triangle	Three	Three
Quadrilateral	Four	Four
Pentagon	Five	Five
Hexagon	Six	Six
Octagon	Eight	Eight
Decagon	Ten	Ten

When we deal with polygons, the word "regular" is important. A **regular polygon** is a polygon that has all edges the same size, and the angles at each edge are the same size. A few examples of these are:

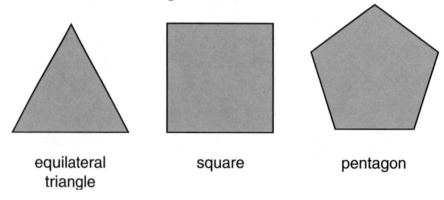

equilateral   square   pentagon
triangle

QUADRILATERALS

Quadrilaterals are, as just defined, four-sided polygons, but there are five special types of quadrilaterals that should be explained.

A **trapezoid** is a quadrilateral that has one pair of opposite legs that are parallel.

A **parallelogram** is a quadrilateral that has both pairs of opposite legs parallel.

A **rhombus** is a quadrilateral that is a parallelogram, but in addition, all four legs are the same length.

A **rectangle** is a quadrilateral that is a parallelogram, but in addition, all four angles are right angles (90°).

A **square** is a quadrilateral that combines one property of a rhombus with another of a rectangle. A square has all four sides that are equal in length, as well as four right angles.

CIRCLES

Since they have no straight sides and no vertices, **circles** have special names for their parts. The edge of the circle is one continuous curved line that goes all the way around the circle. The **radius** of a circle is the distance from the center of the circle to the edge.

The **diameter** of a circle is the distance from one edge of a circle, through the center, and over to the other edge. The diameter is, therefore, two times the size of the radius.

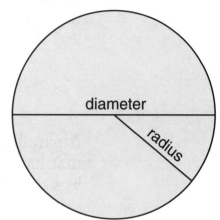

POLYGON EXERCISES

1. Which of the following is a polygon?

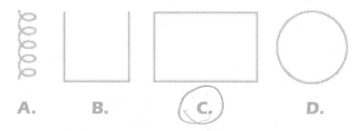

A.   B.   C.   D.

2. If you put two equilateral triangles together, what shape would you form?

   A. A square

   B. A rectangle

   C. A hexagon

   D. A rhombus

3. What polygon has six sides?

(A.) A hexagon

B. An octagon

C. A pentagon

D. A quadrilateral

4. Which shape is *not* a polygon?

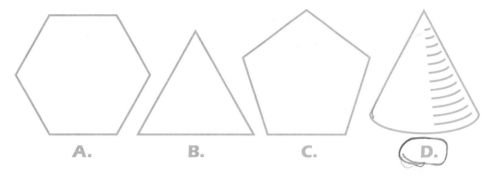

A.   B.   C.   (D.)

5. Tell how many sides each polygon has. Then, name each polygon.

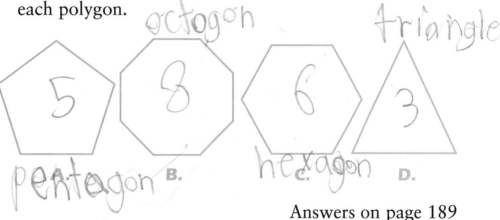

octogon

triangle

5   8   6   3

peAtagon   B.   hexagon   C.   D.

Answers on page 189

SAME SIZE, SAME SHAPE (CONGRUENT FIGURES)

If a polygon is the same size and same shape as another polygon, the two are called **congruent**. A polygon that is congruent to another polygon must be identical to the other one to be considered congruent. That is, corresponding edges must be equal, and corresponding angles must be equal.

Here are two examples of congruent polygons:

If two polygons are in different positions, but are still the same size and shape, they are still congruent.

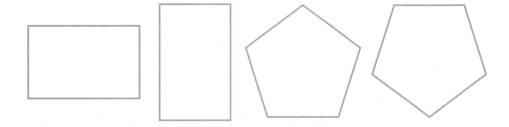

SYMMETRY

If you have a figure that you can cut in half and one half will fold exactly on top of the other half, then that figure is called **symmetric**.

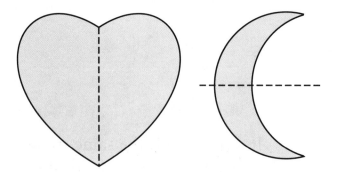

SAME SIZE, SAME SHAPE AND SYMMETRY EXERCISES

1. Which of the following shapes is congruent to this shape:

   A.   B.   C.   D.

2. Which of the following shapes has *only one* line of symmetry?

   A.   B.   C.   D.

3. Choose the two figures that are congruent.

   A.   B.   C.   D.

4. Which letter is *not* symmetric?

   H   Y   R   X

   A.   B.   C.   D.

5. Which figure is congruent to the figure to the right:

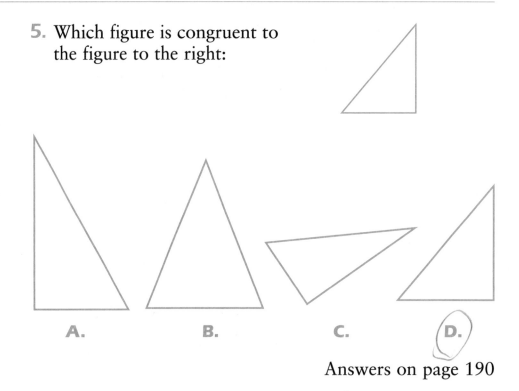

A.   B.   C.   D.

Answers on page 190

SLIDE, FLIP, AND TURN

A figure in two dimensions (a plane figure) can be moved in three different ways.

A **slide** moves a shape a little way across the page. A **turn** spins a shape around a point, usually 90° or 180°. A **flip** turns a shape over, like turning a coin from heads to tails, or a card from front to back. It results in a mirror image of the shape.

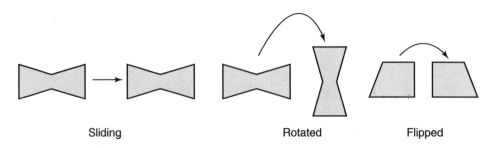

Sliding   Rotated   Flipped

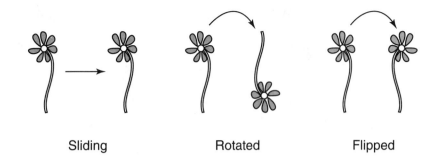

Sliding   Rotated   Flipped

SYMMETRY EXERCISES

1. Which of the following figures represents only a turn?

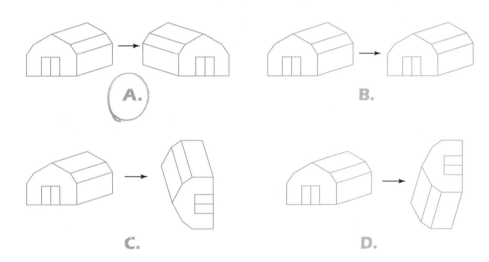

A.   B.

C.   D.

2. Which of the following represents only a slide?

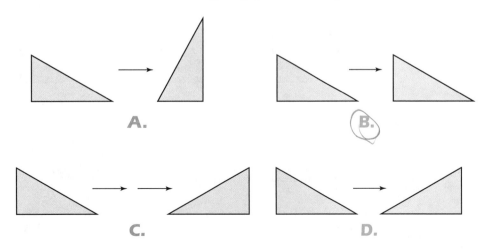

A.   B.

C.   D.

3. Which of the following represents a flip and a turn?

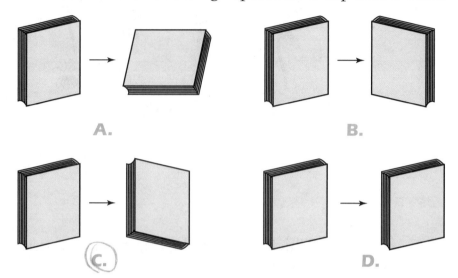

A.   B.

C.   D.

4. Which of the following represents only a flip?

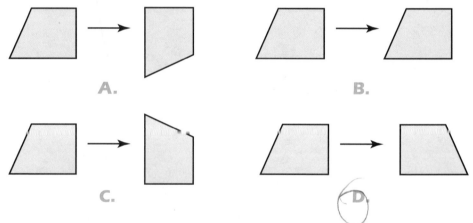

A.   B.

C.   D.

5. Which of the following represents only a rotation?

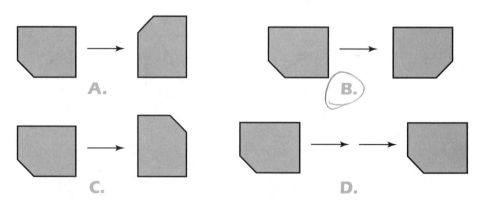

A.   B.

C.   D.

Answers on page 190

COORDINATE GRIDS

The **coordinate grid** is the *x-y* plane, and points on that grid can be located in terms of two points, the *x* point and the *y* point. The *x*-axis runs horizontally, and the *y*-axis runs vertically. When referring to a location on the grid, the *x* point is named first, followed by the *y* point. The resulting location is called an **ordered pair**. An example of this is:

Three points are illustrated here:

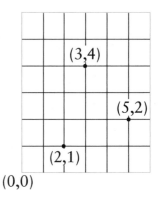

Point (3, 4)
Point (5, 2)
Point (2, 1)

LOCATING POINTS ON THE COORDINATE GRID EXERCISES

Locate these five points on the coordinate grid.

1. (2, 3)
2. (5, 1)
3. (4, 4)
4. (3, 1)
5. (6, 0)

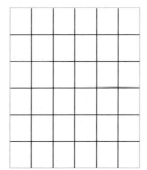

Answers on page 190

MEASUREMENT

We frequently measure items in our lives. You couldn't buy a shirt, shoes, or a jacket without knowing your measurements. The temperature outside on any given day is a measurement that is on everyone's mind, to decide whether to wear that jacket or not.

LENGTH

When we measure length, there are two different systems we use: the metric system and the U.S. standard system.

METRIC SYSTEM

This is a system that uses one standard length, the meter (which is a little longer than a yard in the U.S. system), and uses prefixes on that word to name smaller, or larger, units of length.

> 1 meter (this is the standard in the metric system)
> 10 decimeters = 1 meter
> 100 centimeters = 1 meter
> 1,000 millimeters = 1 meter

Note that we use prefixes to name these lengths:

> The prefix "deci" means one-tenth (and a decimeter is one-tenth of a meter).
>
> The prefix "ceni" means one-hundredth (and a centimeter is one-hundredth of a meter).
>
> The prefix "milli" means one-thousandth (and a millimeter is one-thousandth of a meter).

We can measure larger lengths, as well.

> 1 decameter = 10 meters
> 1 kilometer = 1,000 meters

The prefix "deca" means ten (and a decameter is 10 meters)

The prefix "kilo" means thousand (and a kilometer is 1,000 meters).

Take your metric ruler and measure the diameter of these two coins:

Did you find that the first coin is 15 millimeters in diameter? You could also measure it as 1.5 centimeters.

What was the diameter of the second coin? Is it 30 millimeters? You could also measure it as 3 centimeters.

Now, measure these pencils:

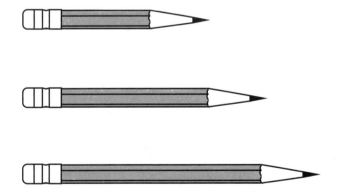

You should have measured the first pencil as 50 millimeters in length. You could also measure it as 5 centimeters.

Did you measure the length of the second pencil as 66 millimeters? You could also measure it as 6.6 centimeters.

What did you find the third pencil's length to be? Was it 8.1 centimeters? It could also be measured as 81 millimeters.

METRIC MEASUREMENT EXERCISES

1. Elaine has four erasers, as shown. Which one is closest to 5 cm?

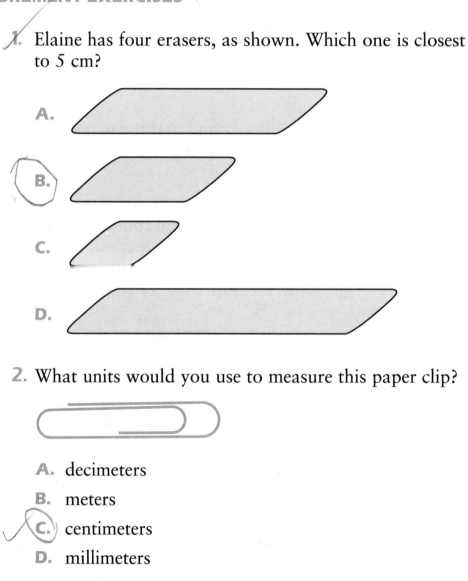

A.

B.

C.

D.

2. What units would you use to measure this paper clip?

A. decimeters

B. meters

C. centimeters

D. millimeters

3. Measure the pen below. What length is it closest to?

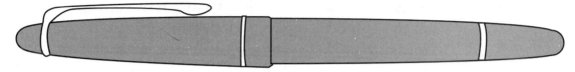

- A. 15 cm
- B. 13 cm
- C. 17 cm
- D. 12 cm

4. What units would you use to measure your school?

- A. decimeters
- B. centimeters
- C. kilometers
- D. meters

5. Measure these two boxes. Pick the pair of numbers that matches their lengths in mm.

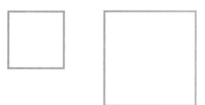

- A. 15, 30
- B. 10, 20
- C. 15, 25
- D. 10, 25

Answers on page 191

CUSTOMARY (U.S. STANDARD) SYSTEM

This is also called the English system because it originated in the United Kingdom. There are four main units of measure in this system:

     1 inch (in.) is this length: ———————
     1 foot (ft) – 12 inches
     1 yard (yd) – 3 feet or 36 inches
     1 mile (mi) – 5,280 feet or 1,760 yards

Use your (customary) ruler to measure these paper clips:

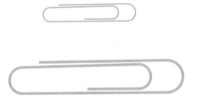

What did you find the length of the small paper clip to be? It should be one inch. Was the larger clip close to 2 inches, but not quite? The second clip should be $1\frac{7}{8}$ inches long.

Try this again, this time on two erasers:

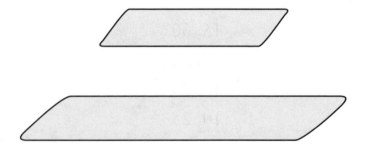

Did you measure the smaller eraser as 2 inches? Was the larger eraser $3\frac{1}{2}$? It should be that length.

CUSTOMARY (U.S. STANDARD) SYSTEM EXERCISES

1. Measure these stamps with your (customary) ruler.
    Which one is between $1\frac{1}{2}$ and 2 inches wide?

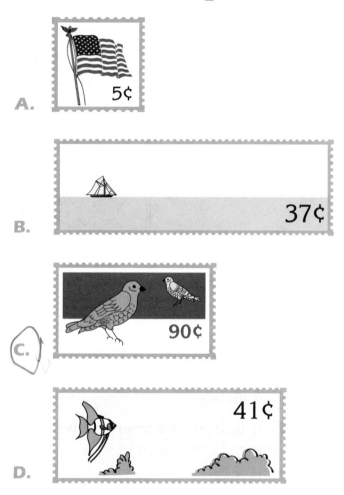

2. Here is an MP3 player. Using your (customary) ruler, tell how long it is.

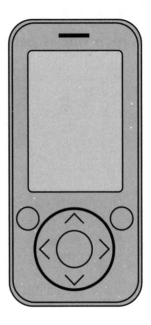

A. 3 inches

B. $2\frac{1}{2}$ inches

C. $3\frac{1}{2}$ inches

D. 4 inches

3. What unit of measurement would we use to measure distance between cities?

A. Inches

B. Feet

C. Yards

D. Miles

4. Measure this ship with a (customary) ruler. Pick the pair of numbers that most closely matches the length and height of the ship.

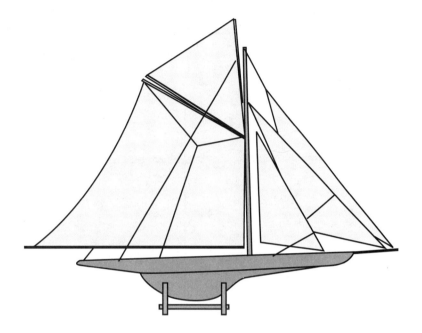

A. 2 inches, 4 inches

B. 3 inches, 5 inches

C. 4 inches, 3 inches

D. 5 inches, 2 inches

5. A notebook is one foot high. How many notebooks will make a yard?

A. 2

B. 3

C. 4

D. 5

Answers on page 191

PERIMETER OF A FIGURE

The **perimeter** of a figure is the total distance around an object. Consider this figure:

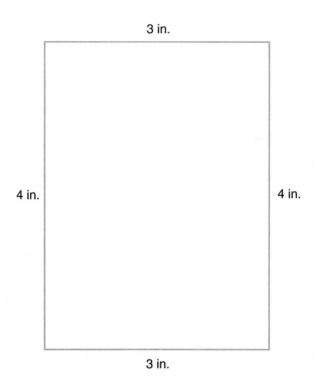

You see here that the perimeter of this rectangle is 3 + 3 + 4 + 4 = 14 inches around. Another way to look at this is to add one side to another and then double it. So (3 + 4) × 2 = 14 inches.

Let's look at another example:

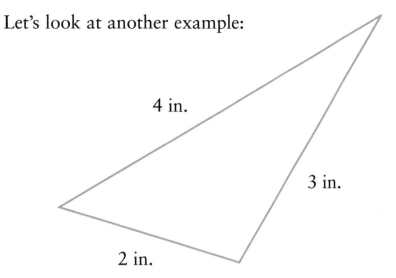

You see here that the perimeter of this triangle is
2 + 3 + 4 = 9 inches around.

Curved figures have a perimeter, as well.

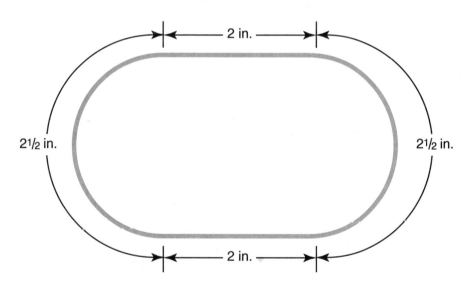

You see here that the top and bottom are 2 in.,
and $2\frac{1}{2}$ in. on each end, so that the perimeter is

$2 + 2\frac{1}{2} + 2 + 2\frac{1}{2} = 9$ in. around.

The perimeter of a circle can be estimated by multiplying
the diameter by 3.

PERIMETER OF SHAPES EXERCISES

1. Sean is putting in a flower bed and wants to put a fence around it. The flower bed is shown below. How many feet of fence should Sean buy?

13 feet

6 feet

A. 38 feet

B. 32 feet

C. 19 feet

D. 78 feet

2. Gilda is decorating the top of a square box, shown below. She wants to put decorative paper around the box top. How much decorative paper does she need?

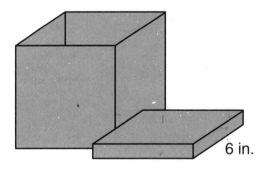

6 in.

A. 36 inches

B. 30 inches

C. 28 inches

D. 24 inches

3. Leroy runs cross country. To practice, he runs around the school track, pictured below. How far around is this track?

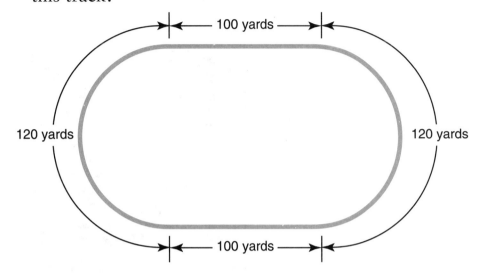

A. 220 yards

B. 340 yards

C. 440 yards

D. 400 yards

4. Andrea bought a pair of shoes. The rectangular box the shoes came in is pictured. What is the perimeter of the box?

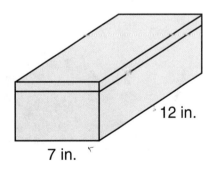

12 in.

7 in.

A. 40 inches

B. 38 inches

C. 34 inches

D. 19 inches

5. George is putting a frame around his favorite poster. The poster is shown below. How much framing material will George be getting?

20 inches

20 inches

A. 400 inches

B. 140 inches

C. 120 inches

D. 80 inches

6. Greg has odd-shaped tables in his restaurant. The tables are shaped like a trapezoid. What is the perimeter of the tables?

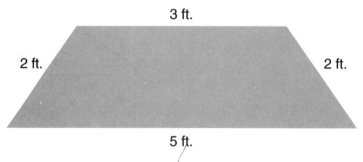

3 ft.

2 ft.   2 ft.

5 ft.

A. 6 feet

B. 10 feet

C. 12 feet

D. 18 feet

7. Joe sells stained glass windows. The one he has on display is a regular hexagon. What is the perimeter of the window?

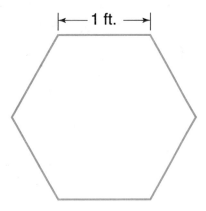

A. 8 feet

B. 6 feet

C. 4 feet

D. 10 feet

Answers on pages 191 to 192

DETERMINE THE AREA OF SHAPES ON A SQUARE GRID

Area is the measure of a two-dimensional shape. It is measured in square units. This is different from the linear measure in earlier sections. To measure area, simply multiply the **length** by the **width**. Some figures, like squares and rectangles, are easy to measure, thus:

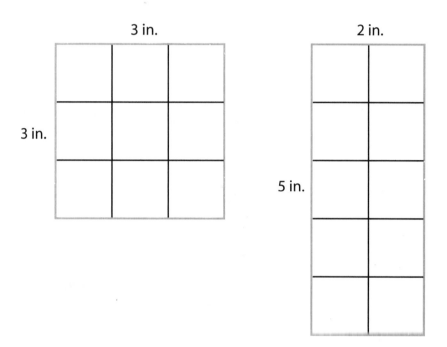

You see that the square is 3 inches on each side; you could count the squares inside to get 9. There is also an equation for the area of the square:

$$A = 3 \times 3 = 9 \text{ square inches,}$$
which is also written as 9 in.^{2}

In similar fashion, the rectangle is 2 inches on one side, 5 inches on the other side, so counting the squares gives you 10. Again, there is an equation for the area of the rectangle:

$$A = 2 \times 5 = 10 \text{ square inches,}$$
which is also written as 10 in.^{2}

You can see, therefore, that square measure is a *different kind* of measure than linear measure. Square measurements always use square units, whether it be square inches (in.^{2}), square feet (ft^2), or square centimeters (cm^2).

Some figures, however, are not as easy to measure. Since area is a measure of the length times the width, how are we to measure a circle, where there are only curved lines?

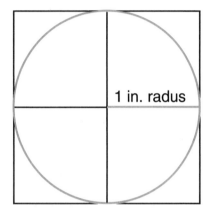

As before, one way of doing this is to draw the figure on a coordinate grid. That way we can count the number of squares inside the figure, and add partial squares to get a reasonable estimate of the area.

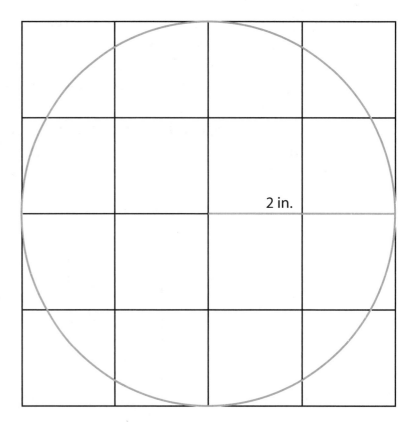

Here's another figure, which is not so difficult to find the area of:

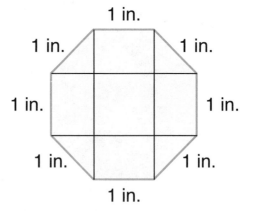

As you can see, with an octagon, it's easy to line up the squares from the grid and get a more accurate measure of the area.

AREA OF SHAPES ON A SQUARE GRID EXERCISES

1. Lynn is having a picnic in the park. She spreads out her blanket, as shown. What is the area it covers?

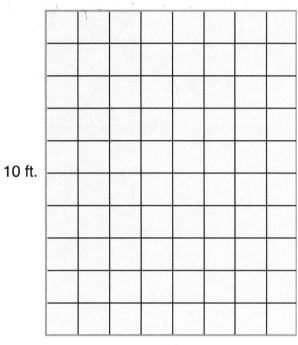

10 ft.

8 ft.

    A. 18 ft^2

    B. 30 ft^2

    C. 60 ft^2

    D. 80 ft^2

2. A wrestling mat is spread out in the gym for practice. What is the area of it?

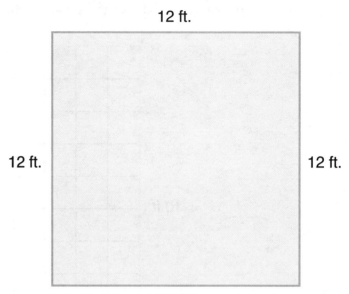

12 ft.

12 ft.   12 ft.

12 ft.

A. 120 ft^2

B. 144 ft^2

C. 164 ft^2

D. 174 ft^2

3. Hattie likes to display her jewelry collection on a felt cloth. What is its area?

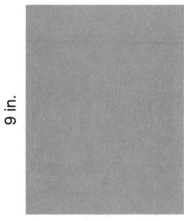

9 in.

7 in.

A. 63 in.^{2}

B. 60 in.^{2}

C. 74 in.^{2}

D. 70 in.^{2}

4. A coffee mug has a radius three inches, as shown. What is its (approximate) area?

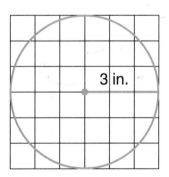

3 in.

A. 20 in.^{2}

B. 24 in.^{2}

C. 28 in.^{2}

D. 30 in.^{2}

5. A small window in the dining room is pictured below. What is its (approximate) area?

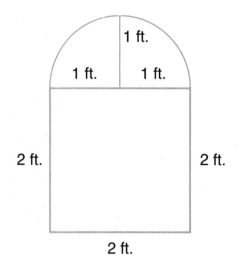

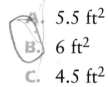
A. 5.5 ft^2

B. 6 ft^2

C. 4.5 ft^2

D. 6.5 ft^2

Answers on page 192

VOLUME

Volume is the amount inside of a three-dimensional figure. It is measured in cubic units. To measure the volume, simply measure the **length** by the **width** by the **height**. As with area, some shapes are easy to calculate:

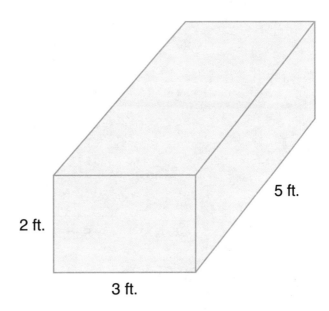

As you can see with this box, the volume is simply $2 \times 3 \times 5 = 30$ cubic feet, sometimes written 30 ft^3.

VOLUME EXERCISES

1. Garrett has a shoebox that is 4 in. by 6 in. by 10 in. What is the volume of Garrett's shoebox?

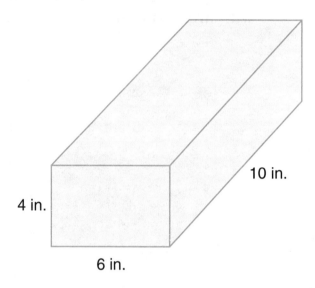

4 in.

6 in.

10 in.

   A. 120 in.^{3}

   B. 200 in.^{3}

   C. 240 in.^{3}

   D. 380 in.^{3}

2. A two-quart container of orange juice is 4 in. by 4 in. by 8 in. (not including the peaked top). What is the inside volume of the container (in cubic inches)?

   A. 128 in.^{3}

   B. 64 in.^{3}

   C. 148 in.^{3}

   D. 132 in.^{3}

3. Alice has a glass prism that looks like this:

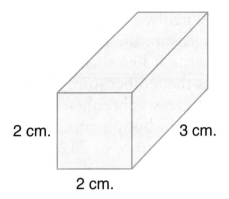

2 cm.   3 cm.

2 cm.

What is the volume of the prism?

A. 14 cm^3

B. 24 cm^3

C. 7 cm^3

D. 12 cm^3

4. Larry bought a refrigerator, and the box it came in is 1 m by 1 m by 3 m. What is the volume of the box?

A. 5 m^3

B. 3 m^3

C. 8 m^3

D. 2 m^3

5. Sylvia bought a can of hairspray that came in a box. The box is 1 inch by 1 inch by 6 inches. What is the volume of the box?

A. 10 in.^3

B. 8 in.^3

C. 6 in.^3

D. 4 in.^3

Answers on page 193

TEMPERATURE

When we talk about how hot or cold it is, we generally measure the **temperature** in degrees **Fahrenheit,** and we write it like this: 32°F. On the Fahrenheit scale, 32 degrees is the temperature at which water freezes into ice. At zero degrees Fahrenheit (0°F), sea water freezes. Most people have a body temperature of 98.6°F, and water boils at a temperature of 212°F.

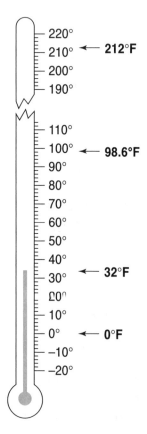

When reading a thermometer, look at the top of the fluid in the glass, and read the numbers next to the top.

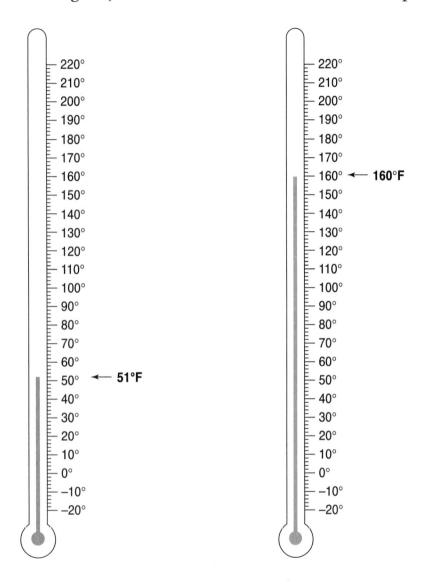

In this figure, the first temperature reading is 51°F, the second reading is 160°F.

The temperature changes all day long. Consider this:

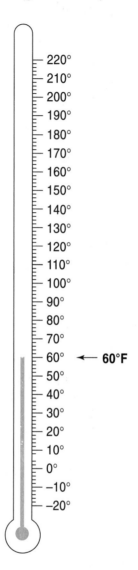

60°F

The reading on this thermometer is 60°F. If the temperature rises four degrees each hour for the next three hours, what would the temperature be?

It would be 72°F because four degrees in three hours is: 4 × 3 = 12, and 60 + 12 = 72.

Let's try another one:

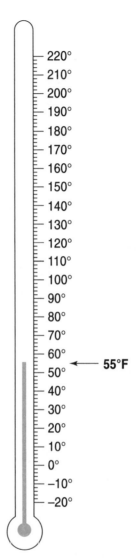

The reading on this thermometer is 55°F. If the temperature drops two degrees every hour for the next four hours, what will the temperature be?

It would be 47°F because two degrees in four hours is: $2 \times 4 = 8$, and $55 - 8 = 47$.

TEMPERATURE EXERCISES

1. Carlos saw that the outside temperature was 45°F at 8 A.M. The temperature rose 2°F per hour until 2 P.M. After that, the temperature fell. What was the highest temperature it made that day?

    A. 59°F

    B. 60°F

    C. 55°F

    D. 57°F

2. Hilary watched the outside temperature closely one night. It was 68°F at 8 P.M. By midnight, it was 40°F. By how many degrees per hour did the temperature fall?

    A. 5°F

    B. 6°F

    C. 8°F

    D. 7°F

3. Jeanne needed to get the temperature of a bath up to 115°F. The bathwater was 82°F, and was rising 3°F per minute. How many minutes will go by before the bath is at the right temperature?

    A. 11 minutes

    B. 12 minutes

    C. 13 minutes

    D. 15 minutes

4. Pat is cooling her sprained ankle. She needs to cool a basin of water down to 35°F. Right now it is 72°F. How many degrees must the water be cooled?

   A. 37°F

   B. 33°F

   C. 31°F

   D. 39°F

5. Amy is an office manager. She keeps the temperature in the office at 71°F. Last night, there was a power failure. When she came in, the temperature was down to 49°F. How far had it fallen from its original temperature?

   A. 12°F

   B. 22°F

   C. 20°F

   D. 21°F

Answers on page 193

TIME

You use time all day, every day. You get up at a certain time each day, go to school for a certain number of hours each day, and usually eat at certain times. **Time** is measured in hours, minutes, and seconds. It's also measured in days, weeks, months, and years, but we won't be talking about those larger units in this section. Hours are measured from midnight with the designation "A.M."; hours from noon have the designation "P.M." Hours are counted 12 to 1, 2, and so forth up to 11. Minutes are designated 0 to 60.

As an example of time measurement, let's look at a digital clock.

In this clock the first number (7) gives the hour. The next two numbers (43) give the minutes. The designation P.M. means that it's after noon, or evening. So, it is 7 hours and 43 minutes after noon.

   Another example looks at a time before noon.

In this clock the first number (8) gives the hour, and the next two numbers (23) give the minutes. A.M. means it is before noon, but it really tells us it is 8 hours and 23 minutes *after midnight*. To find out how far it is before noon, you would need to subtract it from 12:00. Remember, though, that hours go only up to 12, and minutes and seconds go only up to 60. So, 11 − 8 = 3, and 60 − 23 = 37. This clock tells us that we're 3 hours and 37 minutes before noon.

TIME EXERCISES

1. Matthew woke up and looked at his clock. It read 7:34 A.M. He took a shower, dressed, and went down to make breakfast. When he looked at the clock again, it read 8:46 A.M. How much time had gone by since Matthew had gotten up?

   A. 1 hour, 15 minutes

   B. 2 hours, 20 minutes

   C. 1 hour, 12 minutes

   D. 1 hour, 10 minutes

2. Darlene was working on her homework. She looked at the clock, and it read 8:27 P.M. After studying and writing out some homework sheets, she looked at the clock again, and it read 9:38 P.M. How much time had Diana worked on her homework?

   A. 1 hour, 35 minutes

   B. 1 hour, 11 minutes

   C. 1 hour, 27 minutes

   D. 1 hour, 12 minutes

3. Ted coaches wrestling and was giving his wrestlers some drills. He started them at 3:15 P.M. and drilled them for 23 minutes. What time was it then?

   A. 3:23 P.M.

   B. 3:33 P.M.

   C. 3:48 P.M.

   D. 3:38 P.M.

4. Dan was building a bridge along with seven other boys. They worked together $3\frac{1}{2}$ hours, and they finished the bridge. If you were to spread the hours out, how many hours would that be?

   A. 28 hours

   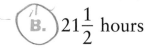 B. $21\frac{1}{2}$ hours

   C. 23 hours

   D. 20 hours

5. Joe was studying for his math exam. He started at 7:15 P.M., and studied 2 hours and 35 minutes. What time was it then?

   A. 9:45 P.M.

   B. 9:40 P.M.

   C. 9:50 P.M.

   D. 9:55 P.M.

Answers on page 193

OPEN-ENDED QUESTIONS

Open-ended questions require you to give an answer and then to explain how you got the answer. You could explain your answer using a chart or graph, using pictures, or using words. The examiners of the test award points for the open-ended questions on a scale of 0 to 3. They award more points for the more complete explanation. They can award only 1 point to a student who gives the correct answer with no explanation.

SAMPLE OPEN-ENDED QUESTIONS
(FOR GEOMETRY AND MEASUREMENT)

1. Harry is painting the floor of a closet in his house. The dimensions of the floor are shown below.

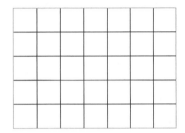

Each square represents a square foot. How would you find out the area of this floor?

Answer to question 1. You could count the number of squares: 35. Another way to solve this would be to multiply the number of feet on the length, which is 7, by the number of feet in the width, which is 5. 7 × 5 = 35. A third way might be to add columns of five, seven times, or to add rows of seven, five times. This third way is really multiplication, of course, but it sometimes is used by students who are not completely comfortable multiplying.

2. Look at the figures below.

 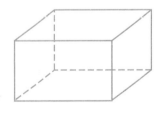

Name the two figures.
Tell how many edges each figure has.
Tell how many faces each figure has.
Tell one way each figure is the same.
Tell one way each figure is different.

Answer to question 2: There are five questions here. The answers to the first question is: The first figure is a three-sided pyramid, and the second figure is a rectangular prism.

The second question asks how many edges each figure has. The pyramid has six edges, and the rectangular prism has twelve edges.

The third question asks how many faces each figure has. The pyramid has four faces, and the rectangular prism has six faces.

The fourth question asks one way in which the figures are the same. There are several ways that the figures are the same:

a. Both figures have only straight edges (no curves).

b. Both figures have only flat faces (again, no curving lines).

c. Both figures are three-dimensional.

The fifth question asks how the two figures are different. There are several ways in which the figures are different:

a. The pyramid is smaller than the rectangular prism.

b. The pyramid has a smaller number of edges than the rectangular prism.

c. The pyramid has a smaller number of faces than the rectangular prism.

d. The pyramid is made of triangles, whereas the rectangular prism is made of rectangles and squares.

e. For the pyramid, the angles made by the edges are all acute angles, whereas for the rectangular prism the angles made by the edges are all right angles.

3. Effram visited a house of worship in a small village. The building is pictured below.

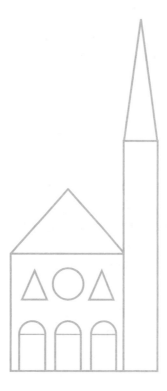

In this house of worship, identify all the kinds of geometric figures you can.

Tell how many of each kind of figure there is.

Tell which figures are congruent to each other.

Answer to question 3: There are three questions here. The answer to the first question is that there are squares, rectangles, triangles, circles, and half-circles in this picture. The answer to the second question is that there is one square, four rectangles, one circle, four triangles, and three half-circles in the picture. To answer the third question, we note that the three doors are congruent rectangles. Likewise, the half-circle windows above each door are congruent to each other. Finally, the two triangle windows on either side of the circular window are congruent to each other.

4. Ted is planning on making a deck in the back of his house. The deck will be 54 feet around the edge, and will be 15 feet long. Tell how wide the deck will be. Ted wants to put little lights every 3 feet around the edge of the deck. How many will he need? How did you get your answer?

Answer to question 4. Since the perimeter is 54 feet, and the length is 15 feet, the two sides of the length are 2 × 15 = 30 feet, and 54 − 30 = 24 so the two sides of the width are 24 feet and 24 ÷ 2 = 12, so the width is 12 feet.

To determine the number of lights around the edge, note that, along the length, there will be a light at the corner, and one every 3 feet, over to the second corner. Marking those lights, they are at 0, 3, 6, 9, 12, and 15 feet. So that makes 6 lights on each length. Since there are two lengths, there are 12 lights needed for the lengths. Now we need to place lights along the width. We already have lights at each corner, so we only need to place lights at the feet markers of 3, 6, and 9 (because at 0 and 12 there are already lights). That makes 3 lights along each width. There are two widths, so we need a total of 6 more lights for the two widths. So the total number of lights we need along the edge of the deck is 12 + 6 = 18 lights.

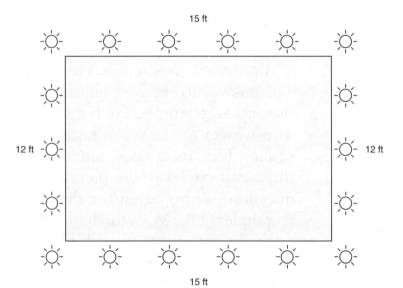

5. How many cubes are needed to make this figure?

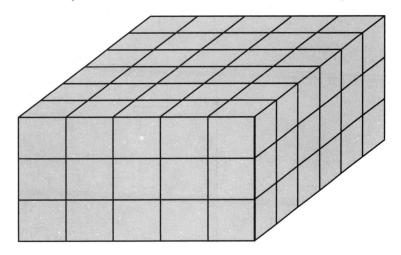

Answer to number 5. Note that the prism is 6 cubes long and 5 cubes wide. So each layer is 6 × 5 = 30 cubes. Now, the prism is 3 cubes high, so we need to multiply 30 by 3: 30 × 3 = 90. The prism requires 90 cubes.

OPEN-ENDED QUESTION (GEOMETRY AND MEASUREMENT) EXERCISES

1. Look at this figure in the grid:

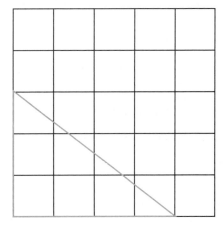

Using the grid, estimate how big the triangle is. How did you arrive at the answer?

2. Below is a plan of Gerry's new dollhouse he's building. Using the paper clip below, estimate the perimeter of the house.

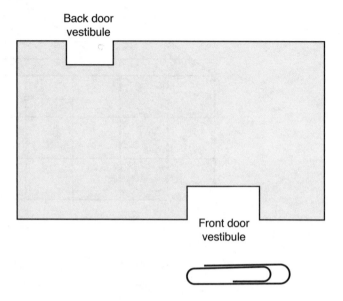

Back door vestibule

Front door vestibule

3. Look at the following seven figures.

Choose all the figures that have just four lines of symmetry, and draw them with those lines of symmetry. Do not choose a figure if it has more than four lines of symmetry. Explain why they are symmetric.

4. Look at the ship below.

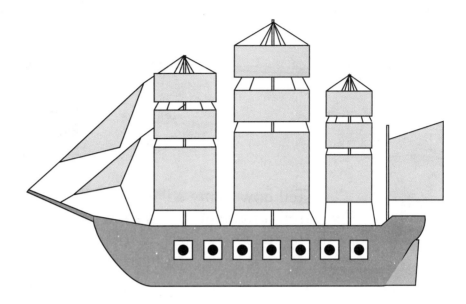

List and name all the geometric figures.
Tell which figures are equivalent to others.
Explain why they are equivalent.

5. Consider the two figures below.

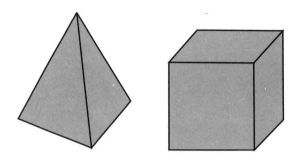

Tell how many edges each figure has.
Tell how many faces each figure has.
Tell one way each figure is the same.
Tell one way each figure is different.

Answers on pages 194 to 195

Chapter 4

PATTERNS AND ALGEBRA (FUNCTIONS)

PATTERNS

Patterns are all around us. Many of the nursery rhymes that you are familiar with describe patterns.

> Hickory, dickory, dock.
>
> A mouse ran up the clock.
>
> The clock struck one;
>
> And down he run.
>
> Hickory, dickory dock.

It has been suggested that the words hickory, dickory, and dock stand for "eight," "nine," and "ten," respectively. In the schoolyard verse for choosing sides, "eeny, meeny, miny, mo" most likely stand for: "one, two, three, four."

The familiar nursery rhyme:

> As I was going to St. Ives,
>
> I met a man with seven wives,
>
> And every wife had seven sacks,
>
> And every sack had seven cats,

And every cat had seven kits.

Kits, cats, sacks, wives,

How many were going to St. Ives.

. . . simply suggests a multiplication problem: $7 \times 7 \times 7 \times 7$.

ACTIVITY: LET'S EXPLORE A PATTERN

Get out a piece of paper. Write 7 up at the top. Add 5 to the 7 and put the result below the 7. Add 5 to the result and put the result below. Continue to do this until you get to around 120. Now, look at the sequence of numbers.

Do you notice any patterns? Are there any patterns in the units place? In the tens place? Add two successive numbers. Do you see a pattern?

Mathematics is full of patterns. Some patterns are easy to see, such as the even numbers (2, 4, 6, 8, etc.). Some are not so easy to see, such as the Fibonnacci sequence (1, 1, 2, 3, 5, 8, 13, 21, etc.). In case you didn't see it; in the Fibonnacci sequence, the next number is found by adding the previous two numbers together (try it with the numbers of the sequence yourself). Mathematics is considered to be the search for patterns in our world. Formulas, such as the area of a rectangle ($A = L \times W$), or area is length times width, is nothing more than a pattern for finding how big a rectangle is. Using patterns, we can find out how big a wall is, or how big a room is (volume). Using patterns, we can predict (to the second) when the sun will rise and when it will set.

Patterns are found in shapes, as well. Look at the array:

What would you expect to see next?

You would expect to see another diamond because the pattern is two diamonds, a square, then two more diamonds.

Sometimes, patterns can be used for box shading. Look at the following rectangles:

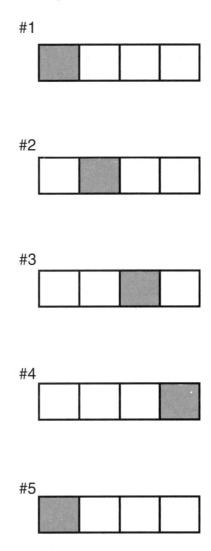

#1

#2

#3

#4

#5

What would you expect the sixth box to look like?

You would expect the sixth box to have the second division shaded because the shaded box goes from the first to the second to the third to the fourth, and back to the first again.

Patterns can also use size. Look at this array:

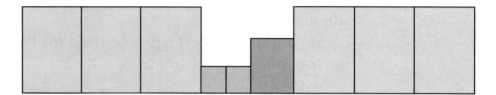

What size would the next two figures be?

The next two figures would be two small squares, following the pattern of three large, two small, and one medium square.

Patterns can combine colors and shapes. Consider the following string of beads

What would you expect the next three beads to be?

The next three beads would be blue round beads because that follows the pattern.

Number arrays can have patterns. The array of numbers

$$2, 4, 6, 8, 10, 12, \ldots$$

is probably familiar to you. It is the even numbers. Now, look at this pattern:

$$1, 3, 5, 7, \ldots$$

What would the next three numbers in this array be?

The next three numbers in this array are 9, 11, and 13 because these are the odd numbers.

You can count down and still find patterns in numbers. Look at this array:

$$55, 51, 47, 43, 39, \ldots$$

What would the next three numbers be?

The next numbers would be 35, 31, and 27 because you are counting down by 4s.

PATTERNS EXERCISES

1. What are the next four numbers in the number array:
5, 8, 11, 14, . . . ?

   17 20 23 26

   Explain the pattern.

   +3

2. What are the next three numbers in the number array:
256, 251, 245, 238, . . . ?

   230 221 211 24

   Explain the pattern.

   By adding 1 to the dec-
   decreasing number

3. What are the next four beads in the string?

   Explain the pattern.

   repeat (2 diamonds, 1 circle, 2 squares, 2 triangles)

4. What are the next three beads in the string?

   blue ⟶

   Explain the pattern.

   2 gray, 3 blue, repeat

5. What are the next three shapes in the array?

Explain the pattern.

repeat (2 moons, 3 stars)

6. What are the next four shapes in the array?

Explain the pattern.

4 squares, 3 triangles, 1 circle repeat

7. Consider these square arrays of dots:

What will the next array be?

Explain the pattern.

1 row & column
2 rows & column
3 rows & column
increase row & column by one

Answers on page 196

FUNCTIONS (MODELING)

A function is a pattern in mathematics that may have one or more steps. There are three ways a function can be shown. Very often it takes the form of a "function machine." Here is an example with one step:

Theodora has a magic machine. When she puts 5 in she gets 9. When she puts 8 in, she gets 12 out. If she puts 14 in, what will come out?

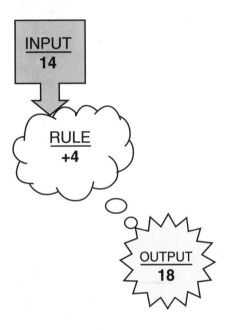

The number 18 will come out. The function adds 4 to the number that goes in.

A second way a function can be shown is with a table:

First number	5	8	14
Second number	9	12	18

A third way functions can be shown is with a formula, or a number sentence:

$$a + 4 = b$$

where a is the first number and b is the second number.

Here's another example of a function: Yu plays a computer game that gives his character strength points every time he goes to a well. When he goes to the well with 6 points, he leaves with 12 points. When he goes to the well with 13 points, he leaves with 26 points. If he goes to the well with 21 points, how many points will he leave with?

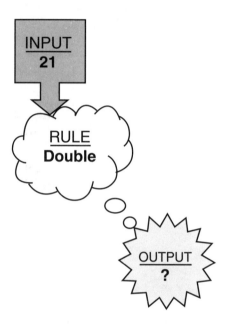

The number 42 will come out. The function machine doubles the number every time.

The table of Yu's strength points would look like this:

Original strength points	6	13	21
Final strength points	12	26	42

The number sentence for this function would look like this:

$$2a = b$$

where a is the original strength points and b is the final strength points.

FUNCTION EXERCISES

1. Guy had a savings account in the bank. He went every week to take some money out. He had $45 to start out with. The first week he had $39 after he took money out. The second week he had $33 after he took money out. If he follows the same pattern, how much money will he have on the third week after he takes out money?

2. Mysterio the Magician has a magic machine. When he puts in 24, he gets back 12. When he puts 18 in, he gets back 9. When he puts 14 in, he gets back 7. If he puts 10 in, what will he get back?

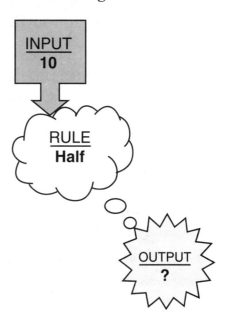

3. The Everwet Pool company is fixing a swimming pool. They need to drain the pool of all its water. The pool originally had 60 inches of water in it. Each hour, the water level went down according to this table:

Hour	0	1	2	3	4
Level	60	52	44	36	?

What will the level be in the fourth hour?

4. Dan is reading a book. On the first day he read up to page 13. On the second day he read up to page 26. On the third day he read up to page 39. What page will he be up to on the fourth day?

5. Renee measured the temperature one day. She charted the temperature on a graph, as shown below:

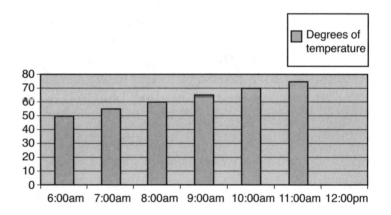

What was the temperature at 12 o'clock?

Answers on page 197

PROCEDURES

Often problems can be solved using procedures. Consider the following problem: Marisol had $9. She wanted to go to the movies, so her mother gave her some more money, and then she had $15. How much did her mother give her?

In this problem, an open sentence would help you solve it. The sentence would look like this:

$$9 + \underline{\quad} = 15$$

because Marisol started out with $9, her mother gave her some money (we don't know how much), and she ended up with $15. To solve this, you would subtract 9 from both sides, and get the answer of $6. Check your answer by putting 6 into the blank in the original sentence, and see that $9 + 6 = 15$ is correct.

Here's another problem using an open sentence: On a hot day in July, Justin wanted to treat his friends to some ice pops. He had four friends with him, so he got five pops (four for his friends and one for himself). Then some more friends came along, and like a good sport Justin got more pops. He then handed around the pops he found that he had handed out eleven, in all. How many friends came along later?

In this problem, use a sentence similar to the first one:

$$5 + \underline{\quad} = 11$$

because Justin started out with five pops, then he got some more, and then he had eleven. We solve this by subtracting 5 from both sides, and get the answer of 6. To check our answer, put 6 in the blank in the original sentence and see that $5 + 6 = 11$ is correct.

Here's another open sentence problem: Tameka had some hairbands. She gave seven to her sister, and then she had sixteen. How many did she have to begin with?

In this problem, again, an open sentence would help you solve the problem:

$$\_\_ - 7 = 16$$

because Tameka had some hairbands (we don't know how many), she gave some to her sister (so we subtract from the unknown amount), and she ends up with 16. To solve this, we would add 7 to both sides, and get the answer of 23. To check our answer, we would put 23 into the open sentence and find that $23 - 7 = 16$ is indeed correct.

PROCEDURE EXERCISES

1. Jon and Tim joined the fire department. One requirement in the physical test is to scale a 12-foot wall. Jon attempted the wall four times, and made it on his fifth attempt. Tim took eight tries and got over on his ninth attempt. How many more times did Tim try than Jon?

2. Brandy has a great pin collection. She went to school with some pins on her hat. A couple of her classmates liked the pins, and bought five from Brandi. When she got home, Brandy counted, and found that she had thirteen pins left on her hat. How many did she start with?

3. Kale works as the cook at the local Emperor of Burgers fast food restaurant. In one hour, he made thirty-two hamburgers. Customers came and bought most of those hamburgers, and then there were three hamburgers left. How many hamburgers did customers buy?

4. Elouise and Harry are great friends from childhood, but they are very competitive. Elouise challenged Harry to a race. They ran for a distance of 100 meters. Over that distance, Elouise took 36 seconds. For the same distance, Harry took 42 seconds. How many more seconds did Harry take?

5. Hassan gets $5 every time he delivers a pizza. He wants to save up $25 for a new bicycle rack. He made three deliveries already today. How many more deliveries must he make to save up what he wants?

Answers on page 197

TEST YOUR SKILLS

1. Samuel cuts grass for his summer job. He's saving up to get a skateboard, which costs $45.00. He saved money over a four-week period as follows:

Week	1	2	3	4
Amount saved	$6.00	$7.50	$9.00	$10.50

If he follows the same pattern, how many weeks will it take for him to save up enough money to get the skateboard?

A. 7

B. 4

C. 6

D. 5

2. Consider this set of beads:

What would the next shape be?

A. Large star

B. Large heart

C. Small heart

D. Large star

3. Robin loves to bicycle. One week, she bicycled the mileage in the following table:

Day	Mon	Tues	Wed	Thur
Mileage	5	6	7	8

If she follows the same pattern, how many miles will she bike on Saturday?

A. 11

B. 10

C. 9

D. 8

4. Consider this number array:

$$32, 45, 58, 71, \ldots$$

What would the next two numbers in the array be?

A. 84, 93

B. 83, 94

C. 83, 96

D. 84, 97

5. Irazarri lives where the winter is very cold with lots of snowfall. She measured the snowfall on Monday, and it was one inch. On Tuesday the snowfall was two inches. On Wednesday the snowfall was three inches. On Thursday she measured the snowfall as four inches. On Friday she went out to measure the snowfall. If it follows the same pattern, how deep will the snowfall be?

   A. Four

   B. Five

   C. Six

   D. Eight

6. When Hilda went to the boardwalk arcade, she found a game she was good at. She won the maximum number of tickets each time she played. She inserted a quarter, and she won 19 tickets. She played a second time, and her total was then 38 tickets. She played a third time, and she had 57 tickets. If she plays a fourth time and gets the maximum number of tickets, how many tickets will she have in all?

   A. 76

   B. 77

   C. 78

   D. 79

7. Consider this number array:

$$32, 64, 128, 256, \ldots$$

   What would the sixth number in this array be?

   A. 512

   B. 2,048

   C. 1,024

   D. 4,096

8. Tim was working at the snack bar on a summer day. When he started the shift, there were two gallons of iced tea in the jug. After a few hours of work, he found that he was making three gallons of iced tea each hour according to the following graph.

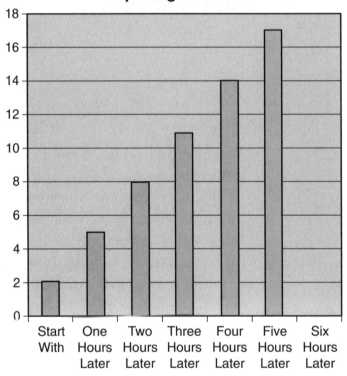

Total Gallons of Tea Made from the Opening of the Stand

How many gallons of iced tea will have been made by the sixth hour (including the 2 to start with)?

A. 14

B. 17

C. 18

D. 20

9. Juanita has her nails done every fourth day. She went to the nail salon on Wednesday. The next time, she went to the salon on Sunday. What day of the week will she go the fifth time she goes to the salon?

   A. Wednesday

   B. Thursday

   C. Friday

   D. Saturday

10. Consider these shapes.

    What would the next circle in the array be?

    A.

    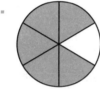

    B.

    C.

    D.

    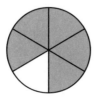

Answers on pages 197 to 199

OPEN-ENDED QUESTIONS

Open-ended questions require you to give an answer and then to explain how you got the answer. You could explain your answer using a chart or graph, using pictures, or using words. The examiners of the test award points for open-ended questions on a scale of 0–3, and they award more points for the more complete explanation. The test rubric awards only 1 point to a student who gives the correct answer with no explanation.

SAMPLE OPEN-ENDED QUESTIONS FOR PATTERNS AND FUNCTIONS

1. Bill plays basketball, and finds that practice the day before a game can help him to get a better score. He kept a chart of the last three games he played and the number of hours he practiced the day before:

Practice (hr)	1	2	3
Points (at the game)	4	8	12

If this pattern stays the same, how many points will he score if he practices five hours?

The student would give the answer of 20 points and then write the linear relationship:

$$y = 4x$$

Other names for the variables could be used, such as:

$$Pr = 4Po$$

Here's another example of an open-ended question:

2. Consider the following number sequence:

$$3, 6, 9, 12, 15, \ldots$$

Give the next five numbers in the sequence.

To answer this, students would need to write down the next five numbers. There are a few ways a student could answer this question. Upon consideration, a student should observe that the numbers go up by threes, so that the next five numbers in the sequence are: 18, 21, 24, 27, and 30. The student would need to write those numbers down, but then he/she would need to explain what the pattern is. The student could tell that this is the three times table, and write down:

$$3 \times 1 = 3$$
$$3 \times 2 = 6$$
$$3 \times 3 = 9 \text{ and so on up to } 3 \times 10 = 30$$

The student might also write down the equation $y = 3x$ (other names for the variables could be used, such as:

$$a = 3 \times b)$$

Another alternative might be for the student to draw a graph of the line:

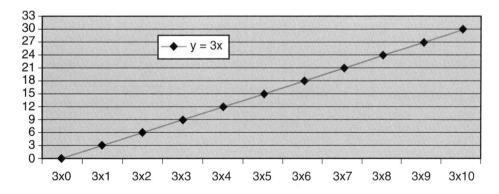

PRACTICE OPEN-ENDED QUESTIONS

1. Roger is very good at washing windows. The neighbors on his street really like his work because he is so good at it, and so he makes a lot of money. He charges by the window. The following table shows how much he earns.

Windows	Amount Charged
2	$3.00
3	$4.50
4	$6.00
5	$7.50
6	$9.00

Tell how much Roger will charge if he does nine windows. Explain how you got your answer.

2. June received a gumball machine when she was six. Now, two years later, the machine doesn't work so well. She puts in a penny, and it gives her two gumballs. She puts in another penny, and she gets another two gumballs. If she has four pennies, how many gumballs will she get, if the gumball machine works the same way? Explain how you got your answer.

3. Frederica was keeping track of the outside temperature over a day for a science project. She measured the temperature each hour until noon, and charted it on the graph:

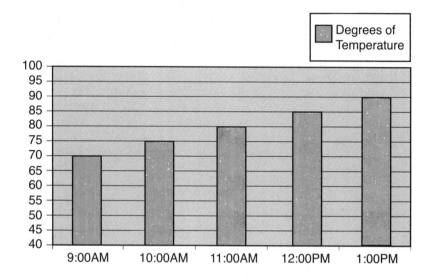

If the temperature follows the same pattern, what will the temperature be at 2 P.M.? Explain your answer.

4. Jiminez has a farm and was driving his hay bailer to bail his hayfield in the afternoon. He noticed that at 2 P.M., his gas tank had 16 gallons. At 3 P.M., his gas tank had 14 gallons. At 4 P.M. his gas tank had 12 gallons. If this trend continues, how many gallons will he have in his tank at 6 P.M.? Explain how you got your answer.

5. Carlita found an old book with a number sequence in it. The book was damaged, and some of the numbers were gone. The sequence that she found looked like this:

400, _____, 360, 340, _____, 300, 280, _____, 240

What are the missing three numbers? Explain how you got your answer.

Answers on pages 199 to 202

Chapter 5

DATA ANALYSIS, PROBABILITY, AND DISCRETE MATHEMATICS

We use data analysis all the time to order our world. Sometimes teachers arrange students in their classes in alphabetical order, and sometimes they order them by height. At the ice cream shop, you can find out how many possibilities you have knowing the ice cream flavors available, the cone types, and the toppings, using discrete mathematics. In this chapter, we'll be looking at these topics.

DATA ANALYSIS

When you collect information, such as the heights of all your classmates, or how many have red shirts and how many have blue shirts, you are collecting **data**. When you put that data in a chart or a graph of some kind and then make judgments based on the graph, that's **data analysis**. Let's look at a simple example of this.

In a class of 22 students, it was found that:

14 students have brown eyes,
6 students have blue eyes, and
2 students have green eyes.

To make this data easier to understand (or see), it can be put in the form of a pictograph:

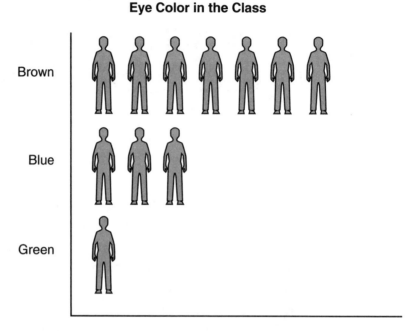

Eye Color in the Class

One student stands for two students

Pictographs often will have pictures stand for the data they represent. The pictures often will be a picture of the thing they represent. So, in this pictograph, there are pictures of students, and each picture of a student stands for two students. In national pictographs, one person might stand for a thousand (or even a million) people.

This data can also be put in the form of a table:

Brown	14
Blue	6
Green	2

The data can also be in the form of a bar graph:

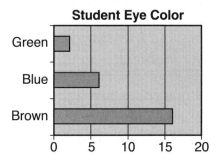

Now, we could go the other way. We could look at a graph and make observations based on that graph. Consider this graph.

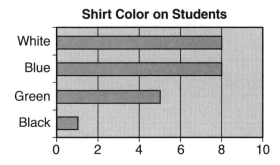

In this example, how many students would you say:

Are wearing black shirts?
Are wearing green shirts?
Are wearing blue shirts?
Are wearing white shirts?

There is one student wearing a black shirt.
There are five students wearing green shirts.
There are eight students wearing blue shirts.
There are eight students wearing white shirts.

We can make certain calculations based on this data, as well. For example:

How many more students are wearing blue than black?

Since there are eight students who are wearing white and one student who is wearing black, there are 8 – 1 = 7. Seven more students who are wearing white than are wearing black.

How many students are in the class?

Since there are eight students wearing white, eight students wearing blue, five students wearing green, and one students wearing black, then, we have

$$8 + 8 + 5 + 1 = 22$$

There are 22 students in the class.

MEASURES OF CENTRAL TENDENCY

There are three measures that help us understand lots of numbers. The **mean** is the numerical average of the numbers. The **median** is the middle number of a group of numbers, and the **mode** is the most frequently appearing number. To understand these measures it's best to illustrate them.

Ferd and his four friends like to collect baseball bobble heads. Ferd has 10, Kyle has 9, Henrietta has 7, Jake has 13, and Geraldine has 7. What are the mean, median, and mode of this collection of bobble heads?

The numbers (sorted from lowest to highest) are: 7, 7, 9, 10, and 13.

The **mean** (average) is calculated by adding up the 5 numbers and dividing by the number of numbers, thus:

$$7 + 7 + 9 + 10 + 13 = 46 \qquad 46 \div 5 = 9.2$$

The average, then, is 9.2.

The **median** is found by sorting the numbers and then taking the middle number. In this case the middle number is 9.

The **mode** is the most frequently encountered number. It's easy to find the mode when they are sorted. The mode is 7.

Let's look at another example:

Sarah has 17 Power Ranger cards to trade with her five friends. Jane has 14, Becca has 25, Mike has 17, Pauline has 15, Fran has 14. What are the mean, median, and mode of these numbers?

Again, sorting the numbers (from lowest to highest), we get:

$$14, \ 14, \ 15, \ 17, \ 17, \ \text{and } 25$$

The **mean** (average) is:

$$14 + 14 + 15 + 17 + 17 + 25 = 102 \quad 102 \div 6 = 17$$

The average, then is 17.

The **median** is the middle number, but we have an even number of numbers here, so what shall we do? When this occurs, take the average of the two middle numbers:

$$15 + 17 = 32 \quad 32 \div 2 = 16$$

The median is 16.

The **mode** is the most frequent number, but we have two most frequent numbers, here. In that case, there are two modes: 14 and 17.

DATA ANALYSIS EXERCISES

1. Emelio was ordering pizza for the class party, and wrote down what all the students wanted. He put the information in the following table:

Plain	X X X X X
Pepperoni	X X X X X X X X X
Mushrooms	X X X
Sausage	X X X X X X X

Based on this information, which of the following is true?

A. There are more students who like sausage than pepperoni.

B. There are fewer students who like mushrooms than plain.

C. There are more students who like plain than sausage.

D. There are more students who like mushroom than sausage.

2. Mrs. Paul, the third grade teacher, asked the students in her class to each make one paper flower for a spring bulletin board she was putting together. She counted the different colored flowers:

> three blue
> five pink
> three purple
> eight white
> four red

If Mrs. Paul was to sort the flowers (from the lowest number to the highest), which color flower would be in the middle of the five different colors?

A. A pink flower

B. A blue flower

C. A red flower

D. A white flower

3. Alice went to the market to get some fruits. She bought

> seven bananas
> six apples
> eight limes
> six lemons
> five pears
> four grapefruits

What was the average (mean) of the fruits?

A. 5

B. 4

C. 8

D. 6

4. Kara drew up a chart that represents the number of students born in the different months of the year:

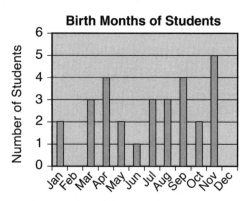

How many students were born in the first half of the year (January to June)?

A. 12

B. 11

C. 14

D. 15

5. Warren is a traffic officer and was looking at the cars driving down Gravel Rd. He wrote his findings in a chart:

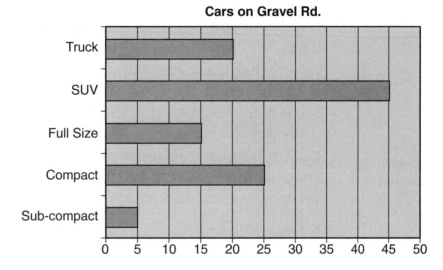

Which statement is true, based on the data?

A. Trucks are the most popular vehicles on Gravel Rd.

B. Compacts are the least popular vehicles on Gravel Rd.

C. Full-size cars are the least popular vehicles on Gravel Rd.

D. The SUVs are the most popular vehicles on Gravel Rd.

Answers on page 202

PROBABILITY

If you look at the chance that something will happen, you are considering the **probability** that it will happen. If a storm comes your way, then there is a good possibility you will be rained on. We say rain is a **likely** event in a storm. If you have a deck of cards, and turn over one card, it is **unlikely** that you'll get the ace of hearts. Getting the ace of hearts is possible, but not probable. There is, in fact, a 1 in 52 chance that you'll turn over an ace of hearts (because there are 52 cards in a deck). The odds are increased, however, if you want an ace of any suit. Since there are four aces in the deck, your chances of turning over an ace is 4 in 52, or if you reduce, 1 in 13. If you roll a die, you can roll one of six numbers: 1, 2, 3, 4, 5, and 6. Thus, you are **certain** to get a 1, 2, 3, 4, 5, or 6, which means that the event will definitely occur. You cannot roll a 20 because 20 is not a number on the face of one die. So, when you roll a die, the probability that you'll roll a 20 is **impossible** (which is an example of an event that cannot happen). To determine the **probability** of rolling a specific number on a die (2), we would determine that there are six faces on a die, and only one has a 2 on it. So the probability of rolling a 2 is 1 in 6, or $\frac{1}{6}$. In a similar way, if we flip a coin, we get a 1 in 2 ($\frac{1}{2}$) chance of the coin landing heads-up.

Consider this spinner:

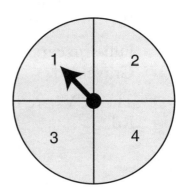

If we spin this spinner, we'll get a 1, a 2, a 3, or a 4. We will not spin a 5 or a 6, or any other higher number. We also will not spin a negative number, or any fractional numbers. The probability of spinning one of those four numbers is 1 in 4, or $\frac{1}{4}$ If we ask the probability of spinning an even number, it is 2 in 4, which becomes $\frac{2}{4}$ or

1 in 2 if you reduce. Sometimes it is written as 1:2. On the other hand, the probability of *not* spinning a 1 is 3 in 4, or $\frac{3}{4}$ In fact, the probability of *not* spinning the 2, the 3, or 4 is the same: 3 in 4, or $\frac{3}{4}$.

PROBABILITY EXERCISES

1. Ellen drew one card from a deck of cards. The probability that she turned over a spade is

   A. 1 in 52

   B. 1 in 13

   C. 1 in 4

   D. 1 in 26

2. Cassie had a bag of marbles. In the bag there were four red marbles, five white marbles, and six yellow marbles. If she drew one marble out of the bag, what is the probability that the marble will be white?

   A. 1 in 5

   B. 1 in 6

   C. 1 in 4

   D. 1 in 3

3. In a game of Candyland, Jennifer spun the spinner, shown:

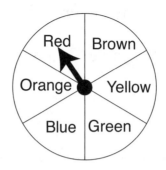

What is the probability that she will get red?

A. 1 in 6

B. 1 in 8

C. 1 in 5

D. 1 in 4

4. Kyle put these letter tiles into a bag:

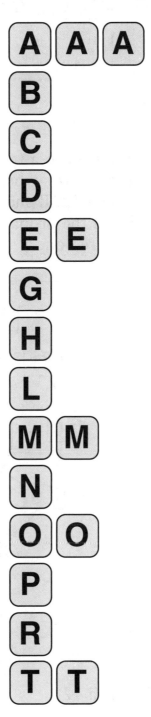

If he drew one out, what is the probability that he would pick an "E"?

A. 1 in 5

B. 1 in 10

C. 1 in 6

D. 1 in 20

5. Janis went to a fair, and got in line for the Ferris wheel. On the Ferris wheel were 12 gondolas, numbered 1 through 12. If she gets into the first available gondola, what is the probability that she will get into an odd numbered one?

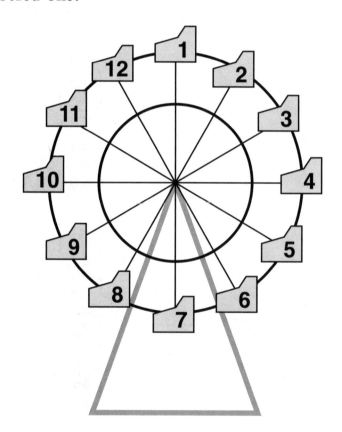

A. 1 in 12

B. 1 in 6

C. 1 in 4

D. 1 in 2

6. Georgina was given the chance to win a trip to Dreamland Amusement Park. She needed to draw a king from a standard, shuffled deck of cards. What is the probability she will draw a king?

   A. 1 in 13

   B. 1 in 52

   C. 1 in 2

   D. 1 in 4

Answers on pages 202 to 203

OPEN-ENDED QUESTIONS: DATA ANALYSIS AND PROBABILITY

1. Ted is on the soccer team and is working out. His record of miles run per day is in the following chart:

Day	Miles
1	2
2	3
3	4

   ▪ Continue the pattern suggested by the table.
   ▪ Or, draw a line graph that shows these figures.

   If he continues in this pattern, how many miles will he run on day 4? How many miles will he run on day 5?

2. Mrs. Kommeth's class went to the dinosaur exhibit at the museum of natural history. At lunch time, everyone took a drink. There are 24 students in her class. The following pie chart shows how many students took what kind of drink.

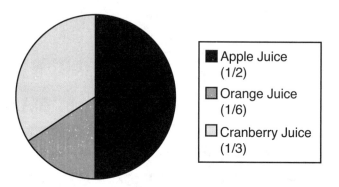

Using the pie chart, and knowing how many students are in the class, find out:

How many students took apple juice? What percentage of the class took apple juice?

How many students took orange juice? What percentage of the class took orange juice?

How many students took cranberry juice? What percentage of the class took cranberry juice?

3. At the shore, Rob's Roasted Peanuts sells double-roasted nuts. He kept track of the number of bags of nuts he sold one day and found the following:

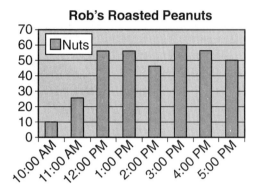

Based on this data, answer the following questions.

A. How many bags of nuts did Rob sell from 10 A.M. to 5 P.M.?

B. How many more bags of nuts did Rob sell at 3 P.M. than at 11 A.M.?

C. Were there any hours in which Rob sold an equal number of bags? Which hours were those? How many bags were sold in each of those hours? How many bags total were sold during those hours?

4. Over at the shore arcade, Gill spins a wheel and gives out stuffed dolphins if a contestant's number comes up. The wheel looks like this:

Take a Chance!

Based on this information, answer these questions.

A. If a contestant places money on 1 and 2, what chances does he have of winning?

B. If a contestant places money on all the odd numbers, what chances does she have of winning?

C. If a contestant places money on only one number, what chances does she have of winning?

Answers on pages 203 to 204

DISCRETE MATHEMATICS: SYSTEMATIC LISTING AND COUNTING

Discrete mathematics involves counting events or things. In the last section, we talked about rolling a die. There are six outcomes when rolling a die: 1, 2, 3, 4, 5, and 6. Counting them is a discrete mathematics task. Flipping a coin gives us two possible outcomes: heads or tails. Picking out a card from a deck of cards involves 52 possible outcomes.

As an example of a discrete mathematics task, think of a girl who is getting dressed for school. She looks in her closet, and sees that there are three blouses that are the right color to go with two skirts there. How many combinations does she have?

To determine the combinations, we match up all possibilities.

You see that six combinations can be made. It is a discrete mathematics task to list all possible combinations. It's important to write them all down so that you can double-check all the combinations. That way, you will avoid counting the same combination twice, or miss a combination.

Another way to solve this is to multiply $3 \times 2 = 6$, although students should show all the combinations.

Let's look at another example. A class of 20 students comes in from the playground to have a snack. They have a choice of an orange and a drink. Some took one or the other; some took both. The teacher counted the snacks left over, and found that 12 students took an orange, and 15 students had a drink. How many *only* had an orange, how many *only* had a drink, and how many had both?

To solve this, we add the students who had an orange to those who had a drink. That is:

$$12 + 15 = 27$$

But, there were only 20 students. So, why do we have 27? The answer is that the extra 7 represents the number of students who had *both* an orange and a drink. To find out how many students had only an orange, subtract 7 from 12 (the number of students who had an orange):

$$12 - 7 = 5$$

This means that 5 students had an orange and no drink.

Next, subtract 7 from 15 (the number of students who had a drink):

$$15 - 7 = 8$$

This means that 8 students had a drink and no orange.

Now, you can analyze this problem using a Venn diagram:

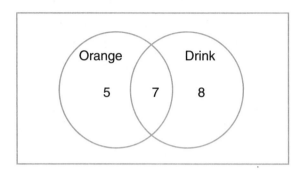

Note that the Venn diagram shows the situation very clearly. It shows the 5 students who had an orange, the 8 students who had a drink, and the 7 students who had both. If you add these three numbers (5 + 7 + 8 = 20), you'll find the original number of students in the class.

DISCRETE MATHEMATICS COUNTING AND LISTING EXERCISES

1. Theodora has three pictures; one of her mom (M), one of her dad (D), and one of her brother (B). She has three different picture frames (1, 2, 3) for the pictures. How many different ways can she put the pictures in the frames?

   A. 9

   B. 4

   C. 6

   D. 8

2. Mr. Kelso, the fire chief of Riverside, has a call chain to call his firefighters, shown below. Who calls firefighter Hill?

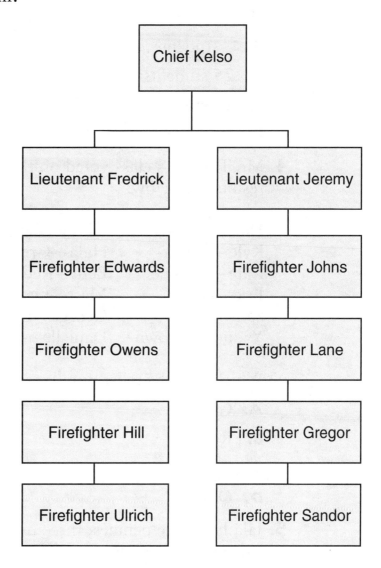

A. Ulrich

B. Lane

C. Gregor

D. Owens

3. In a class of 18 students, all students have either an encyclopedia or a dictionary or both. It was found that 13 have dictionaries home, and 12 have encyclopedias. How many students have both?

   A. 7 students

   B. 25 students

   C. 13 students

   D. 8 students

4. Mr. Thomas's class voted on their favorite color. The results are below:

   Harry—Red   Ginny—Blue   Joe—Red
   Kyle—Green   Hala—Green   Dan—Yellow
   Kate—Orange   Beth—Red   Charlie—Brown
   Eric—Green   Matt—Orange   Pete—Blue
   Sam—Red   George—Blue   Andy—Green
   Yvonne—Brown   Danielle—Red

   Which was the most popular color?

   A. Green

   B. Red

   C. Blue

   D. Orange

5. Jala has three pennies, three nickels, and three dimes in her pocket. How many different combinations can she make if she pulls out two coins?

   A. 6 ways

   B. 8 ways

   C. 9 ways

   D. 12 ways

6. Jules observed pictures in a gallery. There were 14 pictures. Nine pictures had brown frames, and 10 pictures were painted mostly with red paint. How many pictures were in brown frames and painted mostly with red paint?

   A. 9 pictures

   B. 12 pictures

   C. 8 pictures

   D. 5 pictures

Answers on page 205

DISCRETE MATHEMATICS: VERTEX-EDGE GRAPHS AND ALGORITHMS

Discrete mathematics is also concerned with following a set of directions, or finding routes from one place to another.

As an example of finding routes, look at the following diagram:

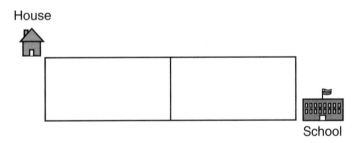

A student living in the house in the upper left-hand corner wants to walk to the school in the lower right hand corner, but likes to take different paths. How many different paths can the student take? There are three different paths the student can take:

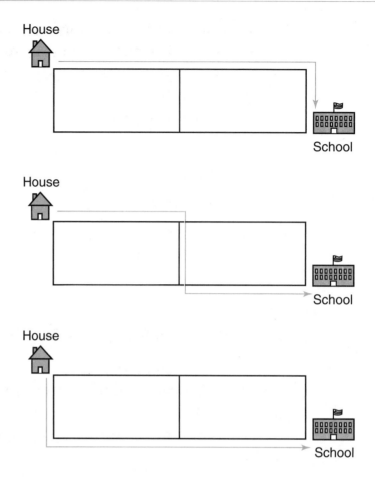

As an example of a set of directions, consider the following map:

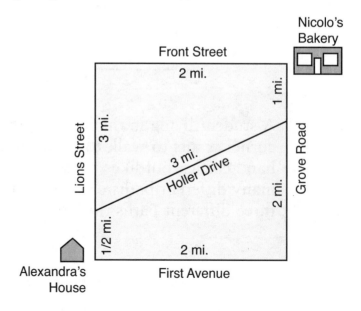

Alexandra works at Nicolo's Bakery and would like to know at least two different ways to get there, in case of a street closure. She is also interested in the shortest way to get to the store.

To answer this, we first consider the two ways to go. One way is to go down Grove Road and across First Avenue. That way is 5 miles. A second way is to go across Front Street and down Lions Street. That way is $5\frac{1}{2}$ miles.

Next, we consider the shortest way for Alexandra. That is to go down Grove Road, across Holler Drive, to Lions Street, and down to her house. That is only $4\frac{1}{2}$ miles.

DISCRETE MATHEMATICS: VERTEX-EDGE GRAPHS AND ALGORITHMS EXERCISES

1. Dierdre wants to visit her cousin, Jenne, four blocks away, as shown below. How many different paths can she take to get to Jenne's house?

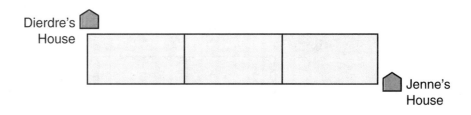

A. Two ways

B. Five ways

C. Three ways

D. Four ways

2. Edward has trumpet lessons at the local community college, and he likes to take different paths. How many different ways can he take to the community college?

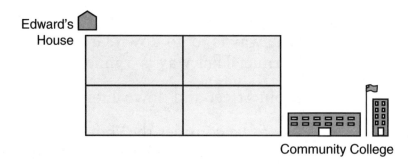

A. Seven ways

B. Six ways

C. Five ways

D. Four ways

3. Leroy runs track and needs to get to the practices from school. The field is four blocks away. How many ways can he go to the track?

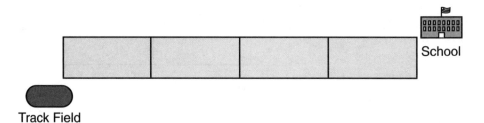

A. Three ways

B. Four ways

C. Five ways

D. Six ways

Answers on page 205

OPEN-ENDED QUESTIONS: DISCRETE MATHEMATICS SYSTEMATIC LISTING, COUNTING, VERTEX-EDGE GRAPHS, AND ALGORITHMS

1. Mac went to the snack machine in the cafeteria for a bag of trail mix. A bag costs 90¢, and the machine will accept nickels, quarters, and dimes. What combination of coins will Mac need to put into the machine to get a bag of trail mix? Show the combination of coins, and then show that the value of the coins adds up to 90¢.

2. At lunchtime, Samantha was comparing what she and her classmates had in their lunchboxes. Her class has 19 students. She found that 11 students had Dandycakes, and 15 (including her) had an orange. How many students had both?

3. Alan needs to pick out a jacket and tie for his recital. He has three jackets and five ties that would fit with each other. How many combinations of outfits can he put together?

4. Ben delivers for Arturo's Pizza Shop. He delivers all over town, but he gets lots of orders from the middle school. Arturo's is five blocks away from the middle school, as shown. Sketch out all the different ways Ben can travel from Arturo's to the middle school.

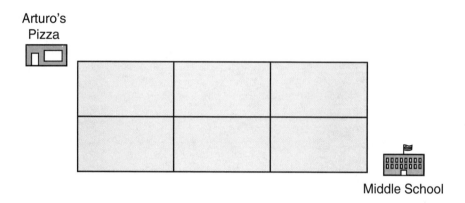

Arturo's Pizza

Middle School

Answers on pages 205 to 208

Chapter 6

PROBLEM SOLVING AND OTHER MATHEMATICAL PROCESSES

PROBLEM SOLVING

You solve problems every day. In fact, you solve problems every hour of every day. When you get up in the morning, you face a problem: "What will I wear?" The answer to that question requires the answer to at least two other questions:

1. What is in my closet that's clean?
2. What is the weather like out today?

This is only one example of many problems you might face in the morning. Depending on your household, you might ask yourself, "What will I eat for breakfast?" or "Will I wash up first, then eat breakfast, or eat breakfast first, then wash up?" Not all people have these problems. Some people will insist on laying their clothes out the night before or following an unchanging routine every morning. But, sometimes a change in routine is unavoidable (if someone is in the bathroom before you, for instance, or you have nothing clean the night before to lay out).

If you go on a school trip, you might consider what to bring (bottled water if you're outside on a hot day or a camera if there are opportunities to take pictures).

At lunch, you might ask yourself whether you want to eat indoors or out.

In the evening, you might consider whether you should do your homework before or after dinner.

A problem is a situation you face where you don't have a clear set of steps to solve it. It can also be a situation that you've never seen before. Facing a situation like that can be scary for many, but there are ways to approach it.

POINTS TO KEEP IN MIND

You are not born with better or worse problem-solving ability than anyone else. Problem solving is a skill, not a talent. All skills are improved by practice, including problem solving. The best problem solvers get better with practice. Do not *ever* believe that you cannot solve problems, because you *can*. You just need to practice at it. There are a few things you should remember as you attempt to solve a problem.

1. Believe that you can do it, because you *can*! If you believe you cannot do it, you won't even try. If you believe you can, you'll try and try and try again until you succeed.
2. Don't think there's only one way to solve a particular problem. Very often there is more than one way to solve a particular problem situation. We'll look at a few strategies here.
3. Keep trying! You can solve it, if you keep at it. If you give up, you certainly won't solve it.
4. Look for the pattern in all of these solutions. Remember that all mathematics is simply looking for patterns, making sure the pattern exists, and attempting to explain the pattern.

METHOD FOR PROBLEM SOLVING

One of the most widely used methods for solving a problem has four steps:

1. Make sense of the problem. Understand all the aspects of the problem. Try to find out all the important information about the problem and to understand what the problem is asking. Think about what information (in the problem) is needed to solve the problem. What information (in the problem) is not needed to solve the problem? This might be the names of the people.

2. Make a plan. Once you understand all the aspects of the problem, you need to select the strategy you will use to solve the problem. Could you draw a picture? Could you make a chart or table? Could you think of a simpler problem? Could you guess the answer and then check it? Could you work backward from the end to the beginning? We'll look at each of these later on.

3. Carry out the plan. Now draw the picture or make up the table or write down the simpler problem or do the guessing and checking, to find the answer to the problem.

4. Check your solution. Now that you have an answer, stop! Look at the answer and think about it. Does it make sense? Does it answer the question completely? This last step is very important, but many people do not do it and give the wrong answer just because they didn't think about it. As an example of this, suppose you had to find out the age of a person. You thought of a plan, carried out the plan, and figured that the person is 900 years old! That of course makes no sense. People new to problem solving often will give an answer that makes no sense just because they don't stop to think about the answer.

NOTE The other four process skills (Communication, Connections, Reasoning, and Representation) are addressed as the various questions are answered. For example, Communication and Representation are addressed when an answer is given, Reasoning goes on in the answering of the question, and Connections are addressed when we look at a problem involving a real world problem, which all of these are.

Here are six problem-solving techniques (lettered A through F):

A. Draw a Picture

If you are trying to tell someone how to get from your home to your school, you might draw a picture of the roads you need to take. Then you can remember what roads to look for and how many turns to make to get there. Very often a picture, or a diagram of some kind, will show the pattern clearly.

Example:

Davy wanted to have an 8-foot long pine board cut into eight pieces. The lumber yard charges $6.00 to cut a board into four pieces (and each cut costs the same). How much will the lumber yard charge Davy?

We'll use the four-step process to solve this problem.

1. Make sense of the problem. We want to know how much Davy will be charged to have a board cut into eight pieces. The useless information is the fact that the wood is pine. We know how much it will cost for the lumber yard to cut it into four pieces. How can we solve this?

2. Make a plan. Drawing a picture would make the number of cuts clearer. That seems to be a good way to proceed. We'll draw a picture of the board and show the number of cuts.

3. Carry out the plan. The board cut into four pieces looks like this:

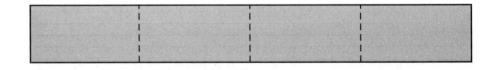

As we see, the board requires three cuts in it to be cut into four pieces. Since the lumber yard charges $6.00 to make three cuts, it is charging $6/3 = $2 per cut. Now, we must extend this to seven cuts.

To cut the board into eight pieces, we use seven cuts, and since each cut is $2, the lumber yard will charge $2 × 7 = $14.

4. Check your solution. Now that we have the answer, let's check to see if it satisfies all the conditions of the problem. Look at the last figure. Is the board cut into eight pieces? Seven cuts gave us eight pieces. Then seven cuts times $2.00 per cut is $14.00. So, it does satisfy the conditions of the problem.

B. Make a Chart or a Table

You have a pocket full of change, and you want to know how much money you have. In this case, a table, where you write down how many quarters, dimes, nickels, and pennies you have, will help you find out how much money you have. Sometimes a chart or table, or diagram of some kind, will reveal the pattern to the solution.

Example:

Grace had pennies, nickels, dimes, and quarters in the pocket of her jumpsuit. If she reached in and pulled out two coins, how many different combinations could she have?

The four-step process.

1. Make sense of the problem. We want to find out how many different combinations of two coins Grace can make with four different kinds of coins. The coins are pennies, nickels, dimes, and quarters. The useless information is that she's wearing a jumpsuit. How can we solve this?

2. Make a plan. If we made a table, we could put down all the combinations. That way, we would avoid duplicate combinations, and make sure all combinations are given. We do not need to know how much each combination is worth in cents. It doesn't ask for that. A table seems to be the best way to proceed.

3. Carry out the plan. The table uses these abbreviations P = Penny, N = Nickel, D = Dime, and Q = Quarter and looks like this:

Coins Used	Combination Number
P P	1
P N	2
P D	3
P Q	4
N N	5
N D	6
N Q	7
D D	8
D Q	9
Q Q	10

So, it would appear that there are 10 combinations of two coins that Grace can make.

4. Check your solution. Do 10 combinations seem reasonable? They do, and more than that, it appears that all the combinations are represented. Note that the table (or chart, if you like) makes it easy to check all the combinations, and see that there are no duplicates. Note, also, that order does not matter. That is, N Q is the same combination as Q N because the question simply asked for how many combinations, not the order of the coins.

C. Think of a Simpler Problem

You go to a party with six other people (besides yourself). How many handshakes does it take for all the people to shake everyone else's hand? Oftentimes a problem has a pattern that can be seen in smaller versions of the problem and can then be built up to the larger problem.

Example:

Gill built a stair with wooden building blocks. They had letters of the alphabet on them. How many blocks does he need to build a stair that has eight steps to it.

The four-step process.

1. Make sense of the problem. We need to find out how many blocks it will take to build a stair with 8 steps to it. The useless information seems to be that the blocks have alphabet letters on them. How can we solve this?
2. Make a plan. A stair with 8 steps seems pretty big. If we look at a smaller stair, we can find the number of blocks easily, and thereby see the pattern that is formed.

3. Carry out the plan. If Gill makes a stair with one step, it will take just one block.

If he makes a stair of 2 steps, it will take 3 blocks.

If he makes a stair of 3 steps, he will take 6 blocks.

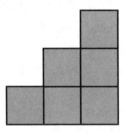

If he makes a stair of 4 steps, he will take 10 blocks.

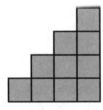

If he makes a stair of 5 steps, he will take 15 blocks.

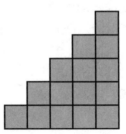

If he makes a stair of 6 steps, he will take 21 blocks.

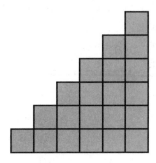

If he makes a stair of 7 steps, he will take 28 blocks.

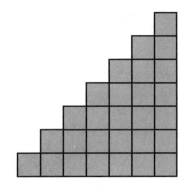

Finally, if he makes a stair of 8 steps, he will take 36 blocks.

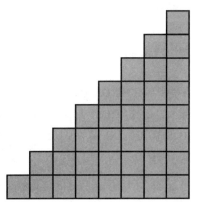

4. Check your solution. Clearly, the figures show that a stair of 8 steps will take 36 blocks. Starting with a simpler problem helps us to build up to the solution. Another way, though, might be to use a table, as with the last problem:

Steps	Blocks
1	1
2	3
3	6
4	10
5	15
6	21
7	28
8	36

Again, this clearly shows how, starting with a simpler problem, the solution to a larger problem can be easily solved.

D. Try Guess and Check

You want to put your collection of 28 model cars into four garages. Can you do it? Sometimes you can get an answer just by guessing it. It usually needs to be checked, though, to see if it fulfills the conditions of the problem.

Example:

Elena has seven friends from ballet class over to her house. She wants to give cookies to each of the seven friends. She has 30 cookies, and she wants to keep two for herself. Can she offer each friend the same number of cookies with none left over?

The four-step process.

1. Make sense of the problem. We want to find out if there is a number of cookies Elena can give her friends with two cookies left over for her. The useless information is that these friends are in ballet class together. What can we do to solve this?

2. Make a plan. One way of solving this is to guess the answer and then check the solution against the conditions of the problem. That seems like a good plan.

3. Carry out the plan. We'll first remember that Elena wants two cookies for herself. That means that she should take away two cookies from the total: 30 − 2 = 28. She needs to distribute 28 cookies to all her friends. Let's try giving the same number of cookies to her friends that she took: two.

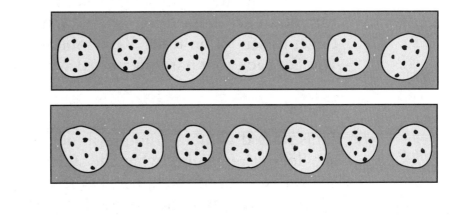

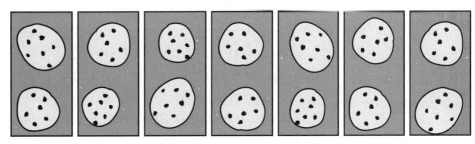

As you can see, if she give two cookies to each person, she'll have 14 cookies left. But that's just the number we used in our first guess. So let's give another two cookies to each friend, like this:

Now, we have it. We have used up all the cookies and each friend has the same number of cookies: four.

4. Check your solution. We see that we have the answer. Each friend can get four cookies with no cookies left over. We guessed the answer, then checked it. It does seem reasonable.

E. Work Backward

You want to plan to get to school on time. School starts at 8:00 A.M. You take 10 minutes to get to school. Breakfast takes 30 minutes to make and eat, and it takes 25 minutes for you to get dressed for school. What time will you start to get to school on time? Remembering each step will help you get back to what you started with.

Example:

Janis started with an amount of money for shopping, but she doesn't remember what she started with. Just before she walked out of the house, her father gave her $20. Then she went out and bought lunch for $7. She bought a belt for $18. While paying for the belt, she met a friend who gave her $15 he owed her. She then bought a blouse for $23, a hat for $15, and she had $7 in the end. How much did she start with?

Here is the four-step process.

1. Make sense of the problem. We want to find out how much she started with at the beginning of her shopping trip. The useless information is the specific things she bought. What will we do to solve this?

2. Make a plan. If we start with the final amount, and work backward, doing the opposite of what Janis did, we'll end up with what she started with. So, for instance, if she spends money, we'll add that amount to the total. If she gets money, we'll subtract it from her total. That way, we'll back up to the original amount. So that seems like a good plan.

3. Carry out the plan. We'll set up a running total, adding or subtracting as needed. Janis ended with $7, so that's what we'll start off with.

She had $7 just before she bought a hat for $15, so

$$\begin{array}{r} \$7 \\ +\$15 \\ \hline \$22 \end{array}$$

She had $22 before she bought the hat. She bought a blouse for $23 before that, so

$$\begin{array}{r} \$22 \\ +\$23 \\ \hline \$45 \end{array}$$

She had $45 before the blouse. She received $15 from a friend before that, so

$$\begin{array}{r} \$45 \\ -\$15 \\ \hline \$30 \end{array}$$

She had $30 before she met her friend. She bought a belt for $18 before that, so

$$\begin{array}{r} \$30 \\ +\$18 \\ \hline \$48 \end{array}$$

She had $48 before she bought the belt. She bought lunch before the belt for $7, so

$$\begin{array}{r} \$48 \\ +\ \$7 \\ \hline \$55 \end{array}$$

She had $55 before she got lunch. She was given $20 before that, so

$$\begin{array}{r} \$55 \\ -\$20 \\ \hline \$35 \end{array}$$

She had $35 before her father gave her $20, so that's what she started with.

4. Check your solution. To check the solution, use the starting amount we found, and do the same operations on it that the problem gives:

$$\$35 + 20 = \$55$$
$$\$55 - \$7 = \$48$$
$$\$48 - \$18 = \$30$$
$$\$30 + \$15 = \$45$$
$$\$45 - \$23 = \$22$$
$$\$22 - \$15 = \$7$$

$7 is what Janis ended up with, so this does check out.

F. Simulate It

That is, use pictures or things to act it out. You have 21 students in your class, and you are going on a field trip. Your school has vans that carry seven students each. How can you divide up the class? Using beans and dividing them up can help you see the solution to this one.

Example:

Anastasia has a collection of 48 small stuffed animals. She wants to put them on her bookcase. The bookcase has lots of shelves, and each shelf holds six stuffed animals. How many shelves does she need to arrange them?

1. Make sense of the problem. We need to figure out how many shelves we will need to arrange the animals. We will be arranging them in groups of six. How can we solve this?
2. Make a plan. If we picture them in groups of six, we can find out how many shelves we need. So drawing a picture, or using beans or some other manipulative, seems like a good way to solve this.
3. Carry out the plan. We draw the picture that arranges the stuffed animals in groups of 6 until all the stuffed animals are used up:

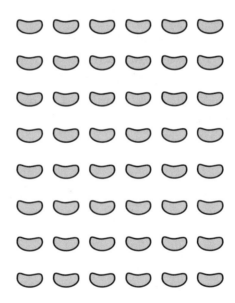

As you can see, we can arrange the stuffed animals on eight shelves of six each. This arrangement will store all the animals, with no partially full shelves. Using mathematical language, this is

$$8 \times 6 = 48$$

4. Check your solution. Did we store all 48 of the stuffed animals? Yes, and there were no shelves that were partially filled with stuffed animals. We used eight shelves to store the stuffed animals six to a shelf. Another way to solve this would be to start with 48 beans, and take 6 away at a time, keeping track of how many times we take 6 away.

PROBLEM-SOLVING EXERCISES

Workspace has been provided for you to use when solving the problems below.

1. Greg is having several friends over for a barbeque in the afternoon. He will be cooking hamburgers, hot dogs, ribs, and chicken drumsticks on the grill. Hamburgers take 20 minutes to cook. Ribs take 35 minutes to cook. Hot dogs take 10 minutes to cook. Chicken drumsticks take 30 minutes to cook. He wants to take all the items off the grill at 3:00 P.M. If he has a grill that will fit all the different meats he is cooking, when should he put each meat on the grill to ensure that they will all come off at 3:00 P.M.?

2. Some members of the Gaskin family (including some aunts and uncles) went to River City Zoo recently. The children invited some of their friends, so that there were more children than adults. They paid $104 to get the whole family in. If the admission price for children is $7, and the admission price for adults is $12, how many children and how many adults went on the Gaskin's zoo trip?

3. Phil and Ben took a kayak trip in Long Lake. They paddled away from Hale Dock at 8:00 A.M., and went east, making 5 mph. They went this way until noon, at which time they pulled onto shore for lunch. Then they got back into their kayaks at 1:00 P.M. and paddled west at a speed of 4 mph, paddling until 4:00 P.M. How close were they to the dock, by then?

4. For Earth Day, Andrew and his brother Henry are picking up plastic bottles and cans along the Passaic River. Andrew picks up six bottles/cans for every five bottles/cans that Henry picks up. They work 2 hours and together they collect 132 bottles/cans. How many bottles/cans did each boy pick up?

5. A triangular array of dots looks like this:

How many dots would there be in a triangular array having 10 dots on each side?

Answers on pages 209 to 212

TEST YOUR SKILLS: PROBLEM SOLVING AND OTHER MATH SKILLS

Use the workspace that has been provided.

1. In a sequence of steps, we know all of them except the first one. What is the first number?

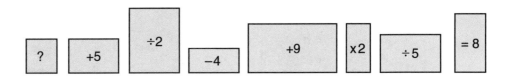

$$? + 5 \div 2 - 4 + 9 \times 2 \div 5 = 8$$

2. Zack wants to put a fence around his mom's vegetable garden. The garden is in one corner of the yard and is shaped like a trapezoid:

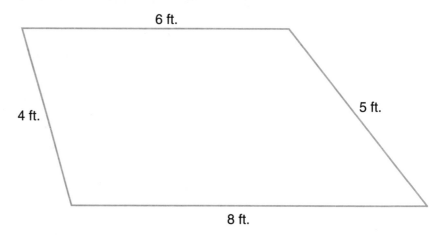

How many feet of fence will Zack need to get?

3. The lockers on the first floor in Third River High School are numbered 100–200. How many lockers have a 6 in them?

4. Consider this number sequence: 23, 28, 26, 31, 29, 34, . . .

   What is the next number in this sequence?

5. Peter wants to plant a tree in his front yard. The yard is 50 feet wide. The dirt ball around the roots of the tree is 4 feet. How far from each side of the yard must he plant the tree to center it in the front yard?

6. Mickey has a quarter, a nickel, and two pennies in his pocket. How many different sums of money can he make?

7. The battleship *New Jersey* gives a break to school groups. For every eight tickets sold, the group gets one free ticket. How many free tickets will a school get if there are fifty in the school group?

8. Consider the following graph:

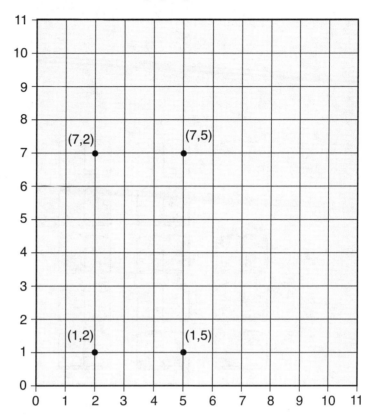

A. If these four points were to be connected, what two-dimensional geometric shape would be formed?

B. What is the perimeter of the shape?

C. If you were to slide this shape four units up, where would the points be plotted on the graph?

9. The following pictograph shows the amount of books sold at Bob and Ray's Book Shop over the last 6 days:

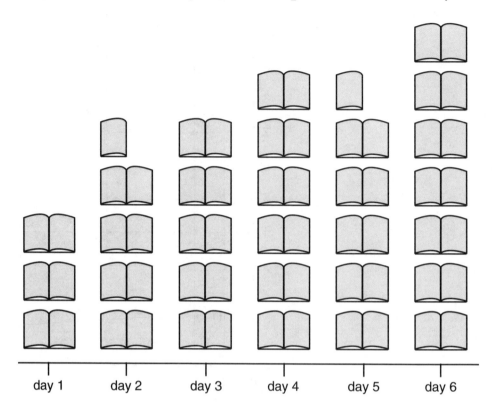

Each book represents 10 books

A. How many books did Bob and Ray's Book Shop sell on the fourth day?

B. On what day did the sales go down?

C. How many books did Bob and Ray's Book Shop sell for the entire six days?

10. Martina is getting outfits together for her Caribbean vacation. She has four blouses (one black, one gray, one blue, and one white), and three pairs of pants (one black, one dark gray, and one white). How many different outfit combinations can she make for herself? Make up a tree or other diagram to show your solution.

If she had two pairs of shoes (one sneakers and one penny loafers), how many blouses, pants, shoes combinations can she make? Use a tree or other diagram to show your solution.

Answers on pages 213 to 222

ANSWERS TO PRACTICE PROBLEMS

CHAPTER 2

ANSWERS TO WHOLE NUMBER PLACE-VALUE EXERCISES

1. **B.** The 0 is the third digit from the right, which is the hundreds position.

2. **C.** The 8 is the first digit from the right, which is the ones position.

3. **D.** The 3 is the fifth digit from the right, which is the ten thousands position.

4. **A.** The 9 is the sixth digit from the right, which is the hundred thousands position.

5. **B.** The 4 is the first digit from the left, which is the hundreds position.

ANSWERS TO DECIMAL PLACE-VALUE EXERCISES

1. **A.** The 0 is the first digit to the left of the decimal, which is the ones position.

2. **C.** The 3 is the second digit to the right of the decimal, which is the hundredths position.

3. **B.** The 9 is the first digit to the right of the decimal, which is the tenths position.

4. **C.** The 4 is the second digit to the right of the decimal, which is the hundredths position.

5. **A.** The 3 is the first digit to the left of the decimal, which is the ones position.

ANSWERS TO COMBINATION PLACE VALUES WITH WHOLE NUMBERS AND DECIMALS EXERCISES

1. **B.** The 7 is the third digit to the left of the decimal, which is the hundreds position.

2. **D.** The 9 is the first digit to the left of the decimal, which is the ones position.

3. **C.** The 0 is the first digit to the right of the decimal, which is the tenths position.

4. **A.** The 0 is the second digit to the right of the decimal, which is the hundredths position.

5. **D.** The 5 is the first digit to the right of the decimal, which is the tenths position.

ANSWERS TO ESTIMATION EXERCISES

1. **C.** 587 is closer to 600 than it is to 500, if you were rounding to the nearest hundred.

2. **D.** 3,971 is closer to 4,000 than it is to 3,000, if you were rounding to the nearest thousand.

3. **B.** 32,800 is the only number that is rounded to the hundreds position. In this number, the 8 is in the hundreds position and every other digit to the right is 0.

4. **B.** $8.00 is the only answer that is rounded to the nearest dollar. Every other answer contains dollars and cents.

5. **B.** $1.66 is closer to $2.00 than it is to $1.00, if you were rounding to the nearest dollar.

ANSWERS TO FRACTION EXERCISES

1. **C.** Five out of eight slices were eaten according to the diagram. The denominator is 8 because that was the total number of pieces in the pizza before any was eaten. The numerator is 5 because it represents the fact that five slices were eaten.

2. **C.** If there are eight pieces of pizza, and five were eaten, that leaves three remaining slices. The denominator is 8 because that was the total number of pieces in the pizza before any was eaten. The numerator is 3 because it represents the fact that three pieces are still remaining.

3. **D.** The denominator in this fraction is 6 because it represents the total number of segments. The numerator is 1 because it represents the value that one segment has on the entire diagram. Hence, one segment represents $\frac{1}{6}$, or one-out-of-six possible.

4. **C.** The denominator in this fraction is 3 because it represents the total number of segments. The numerator is 2 because it represents two segments, which is what the question asked for. Hence, the distance of two segments combined is equal to $\frac{2}{3}$ of the entire line. In other words, it is two-out-of-three possible.

5. C. Because the given denominators are all 5, we know this diagram is split into fifths. In the second missing space, we notice that it is directly to the left of the $\frac{3}{5}$ that the diagram gives us. From this, we can just count backward one and come up with the fraction $\frac{2}{5}$.

The denominator does not change because it is still referring to the total of five segments in the diagram. The numerator changed from 3 to 2 because the distance traveled from the start is now one less, so it needs to be one less.

ANSWERS TO EVEN/ODD EXERCISES

1. B. 45,981 is an odd number because in the smallest place value is a 1, which is an odd number. Therefore, the entire number 45,981 is odd.

2. A. 912,780,312 is an even number because in the smallest place value is a 2, which is an even number. Therefore, the entire number 912,780,312 is even.

3. A. 412 is an even number because in the smallest place value is a 2, which is an even number. Therefore, the entire number 412 is even.

4. C. 501 is an odd number because in the smallest place value is a 1, which is an odd number. Therefore, 501 is an odd number.

5. D. 722 is an even number because in the smallest place value is a 2, which is an even number. Therefore, 722 is an even number.

ANSWERS TO COMPARING NUMBERS EXERCISES

1. **D.** 55,001 is greater than 54,913. Both numbers have a 5 in the ten thousands position. In the thousands position, 55,001 has a 5, which is greater than the 4 in 54,913.

2. **A.** 781,701 is greater than 781,401. Both numbers have a 7 in the hundred thousands position. Both numbers have an 8 in the ten thousands position. Both numbers have a 1 in the thousands position. In the hundreds position, 781,701 has a 7, which is greater than the 4 in 781,401.

3. **A.** The only statement that is true is 412 is larger than 387. If you look at the hundreds position, 412 has a 4, which is greater than the 3 in 387.

4. **A.** 79,654 and 694 are the only even numbers. We know this because they both end with an even digit. So in comparing these two numbers, we know that 79,654 is obviously larger than 694.

5. **C.** 101 and 593 are the only odd numbers. We know this because they both end with an odd digit. So in comparing these two numbers, we know that 593 is greater than 101.

ANSWERS TO ADDITION EXERCISES

1. **C.**

$$\begin{array}{r} {}^{1} \\ 58 \\ +9 \\ \hline 67 \end{array}$$

9 plus 8 is 17. Bring down the 7, carry the 1. 1 plus 5 is 6. So the answer is 67.

2. A.

$$\begin{array}{r} \overset{1}{}\overset{1}{} \\ 8{,}741 \\ +509 \\ \hline 9{,}250 \end{array}$$

9 plus 1 is 10. Bring down the 0, carry the 1. 1 plus 4 plus 0 is 5. 7 plus 5 is 12. So bring down the 2 and carry the 1. 1 plus 8 is 9. So the answer is 9,250.

3. B. Solving these problems individually will yield only one as truthful: 5,413 + 801 = 6,214.

4. A.

$$\begin{array}{r} 1\ 1 \\ 543 \\ 399 \\ +42 \\ \hline 984 \end{array}$$

5. D.

$$\begin{array}{r} 1 \\ 41 \\ 32 \\ 99 \\ +14 \\ \hline 186 \end{array}$$

ANSWERS TO SUBTRACTION EXERCISES

1. B.
$$\begin{array}{r} 23 \\ -11 \\ \hline 12 \end{array}$$

3 minus 1 is 2. 2 minus 1 is 1. So the answer is 12.

2. B.
$$\begin{array}{r} {}^{3}\!\!\!\!\diagdown{}^{1}\!\!43 \\ -\ 9 \\ \hline 34 \end{array}$$

You cannot subtract 9 from 3 so you must borrow from the 4. The 4 becomes a 3 and the 9 becomes original 3 becomes a 13. 13 minus 9 is 4. 3 minus nothing is 3. So the answer is 34.

3. C.
$$\begin{array}{r} 52 \\ -21 \\ \hline 31 \end{array}$$

2 minus 1 is 1. 5 minus 2 is 3. So the answer is 31.

4. D.
$$\begin{array}{r} {}^{4}\!\!\!\!\diagdown{}^{1}\!\!57 \\ -49 \\ \hline 8 \end{array}$$

You cannot subtract 9 from 7 so you must borrow from the 5. The 5 becomes a 4 and the 7 becomes a 17. 17 minus 9 is 8. 4 minus 4 is 0. So the answer is 8.

5. C.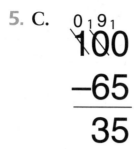

You cannot subtract 5 from 0, so you must move to the left to borrow. There is a second zero there which you cannot borrow from so you must again move to the left to try and "help" that one out so you can borrow from it. Borrowing from the 1 makes it a 0. The middle 0 now becomes a 10. Now that it is a 10, it is able to help that first 0 out. Cross out the 10 and make it a 9, then make the 0 on the right a 10. Now we are ready to do some subtraction. 10 minus 5 is 5. 9 minus 6 is 3. So the answer is 35.

ANSWERS TO MULTIPLICATION FACT EXERCISES

1. **A.** *Hint:* If you count by 5s seven time, you will see the answer is 35.

2. **B.** *Hint:* If you count by 2s nine times, you will see the answer is 18.

3. **A.** If you do the nines trick using your fingers you will see the answer is 63.

4. **D.** Any number times one is that number. So 72×1 is 72.

5. **C.** Any number times zero is zero.

ANSWERS TO DIVISION FACT EXERCISES

1. **B.** In the problem 42 ÷ 6, there are a total of seven 6s in the number 42 with no remainder.

2. **A.** In the problem 25 ÷ 5, there are a total of five 5s in the number 25 with no remainder.

3. **D.** In the problem 63 ÷ 7, there are a total of nine 7s in the number 63 with no remainder.

4. **A.** In the problem 73 ÷ 8, there are a total of nine 8s in the number 73 with one left over so the answer is 9 remainder 1.

5. **B.** In the problem 57 ÷ 9, there are a total of six 9s in the number 57 with three left over so the answer is 6 remainder 3.

CHAPTER 3

ANSWERS TO POLYGON EXERCISES

1. **C.** The rectangle is the only polygon of these choices.

2. **D.** Two equilateral triangles will go together, thus:

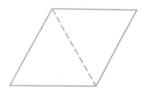

3. **A.** A hexagon. The prefix "hex" means six.

4. **D.** A cone. A cone is a three-dimensional figure, not a two-dimensional figure.

5. **A.** 5, pentagon; **B.** 8, octagon; **C.** 6, hexagon; **D.** 3, triangle

ANSWERS TO SAME SIZE, SAME SHAPE AND SYMMETRY EXERCISES

1. **D.** The pentagon is rotated 180°.

2. **B.** The isosceles triangle has only one line of symmetry.

3. **A** and **D** are congruent.

4. **C.** The letter R is the only letter that is not symmetric.

5. **D.** This right triangle is congruent to the figure.

ANSWERS TO SYMMETRY EXERCISES

1. **C.** The barn is turned.

2. **B.** The triangle is slid.

3. **A.** The book is flipped and turned.

4. **D.** The trapezoid is flipped.

5. **A.** The pentagon is turned.

ANSWERS TO LOCATING POINTS ON THE COORDINATE GRID EXERCISES

1-5.

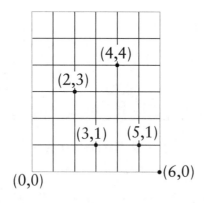

ANSWERS TO METRIC MEASUREMENT EXERCISES

1. **B.** This eraser is 4.5 cm, which is very close to 5 cm.

2. **C.** Centimeters is the best unit to measure paper clips.

3. **A.** The pen is 15 cm long.

4. **D.** Meters is the best unit of measure to use when measuring a building.

5. **C.** The squares are 15 mm and 25 mm long, respectively.

ANSWERS TO CUSTOMARY (U.S. STANDARD) SYSTEM EXERCISES

1. **C.** The stamp is $1\frac{1}{2}$ inches wide, which is closest to $1\frac{3}{4}$ inches.

2. **A.** The MP3 player is 3 inches long.

3. **D.** Cities are miles apart.

4. **C.** The ship is 4 inches long, and 3 inches high.

5. **B.** A yard is 3 feet long.

ANSWERS TO PERIMETER OF SHAPES EXERCISES

1. **A.** The width of the flower bed is 6 feet, and the length is 13 feet. $6 + 13 + 6 + 13 = 38$ feet. Another way to look at this: $(6 + 13) \times 2 = 38$ feet.

2. **D.** The box is 6 inches on a side. So $6 + 6 + 6 + 6 = 24$ inches. Another way of looking at this: $6 \times 4 = 24$ inches.

3. **C.** The straightaways are each 100 yards, and the end half-rounds are 1230 yards, so
   $100 + 120 + 100 + 120 = 440$ yards.
   Another way of looking at this:
   $(100 + 120) \times 2 = 440$ yards.

4. **B.** The shoebox is 7 inches on one side, 12 inches on another side. $7 + 12 + 7 + 12 = 38$ inches. Another way of looking at this: $(7 + 12) \times 2 = 38$ inches.

5. **D.** The poster is 20 inches on a side, so $20 + 20 + 20 + 20 = 80$ inches. Another way of looking at this is $20 \times 4 = 80$ inches.

6. **C.** Adding the four lengths: $3 + 2 + 5 + 2 = 12$ feet, and so the perimeter is 12 feet.

7. **B.** Each side is one foot, so the perimeter is $1 + 1 + 1 + 1 + 1 + 1 = 6$ feet.

ANSWERS TO AREA OF SHAPES ON A SQUARE GRID EXERCISES

1. **D.** $8 \times 10 = 80 \ ft^2$

2. **B.** $12 \times 12 = 144 \ ft^2$

3. **A.** $7 \times 9 = 63 \ in.^2$

4. **C.** Counting the squares, and estimating the squares under the circle, gives $28 \ in.^2$.

5. **A.** Counting the squares, and estimating the squares under the semicircle, gives $5.5 \ ft^2$.

ANSWERS TO VOLUME EXERCISES

1. **C.** 240 in.^{3}
2. **A.** 128 in.^{3}
3. **D.** 12 cm^3
4. **B.** 3 m^3
5. **C.** 6 in.^{3}

ANSWERS TO TEMPERATURE EXERCISES

1. **D.** The temperature rose 2°F each hour for 6 hours, so it rose a total of 12°F, and 45 + 12 = 57°F.

2. **D.** Between 8 P.M. and midnight, the temperature fell 28°F. 28 divided by 4 gives 7, so the temperature dropped 7°F per hour that night.

3. **A.** The bath needed to be warmed up 33°F. Since the temperature rose by 3°F per minute, it will take 11 minutes to get to the right temperature.

4. **A.** 72 − 35 = 37. So the temperature must be dropped 37°F to get to the proper temperature of 35°F.

5. **B.** 71 − 49 = 22. So the temperature had fallen 22°F from the original temperature.

ANSWERS TO TIME EXERCISES

1. **C.** 1 hour, 12 minutes
2. **B.** 1 hour, 11 minutes
3. **D.** 3:38 P.M.
4. **A.** $3\frac{1}{2} \times 8 = 28$
5. **C.** 9:50 P.M.

ANSWERS TO OPEN-ENDED QUESTION (GEOMETRY AND MEASUREMENT) EXERCISES

1. Consider the figure below. If we count the full squares (or almost full squares) in the triangle, we get 5. Then, looking at the partial squares, we find that there are 2, which when added up gives one whole square. Then, the sum of these is 5 + 1 = 6. This is the area of the triangle.

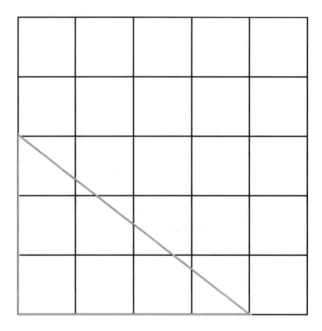

2. Note that the paper clip in the diagram is the same size as a standard (small) paper clip. If we place the paper clip along the perimeter of the dollhouse, we will get a perimeter 8 paper clips long.

3. There are just three of these figures that have four lines of symmetry: the square, the cross, and the ink blotch.

4. There are two triangles (the jibs). There are six rectangles (the top two sails on each of the three masts). There are ten squares (the bottom sails on all three masts, and the seven cannon ports). There are seven circles (the seven cannons). There is one trapezoid (the spanker sail).

   The seven squares that are the cannon ports are all equivalent to each other. The seven circles that are the cannons are all equivalent.

   The top two rectangles on each mast are equivalent because they have corresponding legs that are equal, and the angles are equal because they are all right angles. The squares that are the cannon ports are all equivalent because their sides all have the same length, and the angles are all right angles. The cannon circles are all equivalent because they all have the same radius.

5. The pyramid has eight edges. The cube has twelve edges.

   The pyramid has five faces. The cube has six faces.

   Similarities: The pyramid and the cube are both approximately the same size. The pyramid and the cube both have only straight edges. The pyramid and the cube both have only flat faces.

   Differences: The pyramid and the cube each have a different number of faces. The pyramid and the cube each have a different number of edges.

CHAPTER 4

ANSWERS TO PATTERNS EXERCISES

1. The next four numbers in the array would be 17, 20, 23, and 26. The pattern is that the next number in the array is increased by 3.

2. The next three numbers in the array would be 230, 221, and 211. The pattern is that the next number in the array is decreased by one more each time (The difference between the first and the second number is 5. The difference between the second and third numbers is 6, the difference between the third and fourth numbers is 7, and so forth.)

3. The next four beads would be two blue rhombus-shaped beads, one white circle, and one gray square. The pattern is two blue rhombus-shaped beads, one white round bead, two gray square beads, and two blue triangular beads, then the pattern repeats.

4. The next three beads would be 3 blue circles. The pattern is two gray round beads, three blue round beads, and 2 gray round beads, and then the pattern repeats.

5. The next three shapes would be three small stars. The pattern is two large moons, three small stars, and then the pattern repeats.

6. The next four shapes would be three triangles and one circle. The pattern is four boxes, three triangles, one circle, and then the pattern repeats.

7. The next array would be a square array of dots, four on a side. The pattern is square arrays of dots, first one, then two, then three, then four.

ANSWERS TO FUNCTION EXERCISES

1. On the third week, Guy will have $27 in the bank. The function subtracts 6 each time: $a - 6 = b$.

2. Mysterio's machine divides the input number by 2 each and every time, so the number that comes out will be 5 if he puts 10 in.

3. The level in the pool will be 28 inches. The pool is drained 8 inches of water each hour.

4. Dan will be up to page 52. He reads 13 pages per day.

5. Renee measured the temperature at 80 at 12 o'clock. The temperature goes up 5 degrees each hour.

ANSWERS TO PROCEDURE EXERCISES

1. $9 - 5 = \_\_ \rightarrow 9 - 5 = 4$. Tim tried 4 times more than Jon.

2. $\_\_ - 5 = 13 \rightarrow \_\_ = 13 + 5 = 18$. Brandy had 18 pins on her hat in the beginning.

3. $32 - \_\_ = 3 \rightarrow 32 - 29 = 3$. Customers bought 29 hamburgers in that hour.

4. $42 - 36 = \_\_ \rightarrow 42 - 36 = 6$. It took Harry 6 more seconds to run the 100 meters.

5. $15 + \_\_ = 25 \rightarrow 15 + 10 = 25$. Hassan needs to make two more deliveries to make his goal.

ANSWERS TO TEST YOUR SKILLS

1. **D.** The table shows that he saves $1.50 more each week than the previous week. In week five he would save $12.00. If we add the first five weeks together, we get

$$6.00 + 7.50 + 9.00 + 10.50 + 12 = 45$$

So Samuel can get the skateboard when he gets his money in week five.

2. **B.** Following the pattern, after three small hearts comes two large hearts.

3. **B.** Following the same pattern means that she bikes one more mile each day, and since she biked 8 miles on Thursday, she will bike 10 miles on Saturday.

4. **D.** The next number in the array is increased by 13 from the previous number: $71 + 13 = 84$ and $84 + 13 = 97$.

5. **B.** Following the pattern, the snowfall should be five inches on Friday, because it goes up one inch per day.

6. **A.** The number goes up by 19 each time, and $57 + 19 = 76$

7. **C.** The number is doubled each time, and the fourth number is 256. That makes the fifth number 256×2 or 512, and the sixth number in the array is 512×2 or 1,024.

8. **D.** The jug starts out at 2 gallons, and 3 are made each hour. So, six hours into the shift, Tim has made 3×6 or 18 gallons, plus the 2 that he started with, gives 20.

9. **C.** If we look at the pattern, she will go to the salon on Friday for the fifth time:

Wednesday   to the salon (first time)
Thursday
Friday
Saturday
Sunday   to the salon (second time)
Monday
Tuesday

Wednesday
Thursday   to the salon (third time)
Friday
Saturday
Sunday
Monday   to the salon (fourth time)
Tuesday
Wednesday
Thursday
Friday   to the salon (fifth time)

10. **A.** Following the pattern, the shading must be A.

ANSWERS TO PRACTICE OPEN-ENDED QUESTIONS

1. Roger will charge $13.50 for washing nine windows. There are a few ways to explain this.

    One way is to simply extend the table. Looking at the table, you see that the price goes up by $1.50 for each window. If that pattern is extended, the table looks like this:

Windows	Amount Charged
2	$3.00
3	$4.50
4	$6.00
5	$7.50
6	$9.00
7	$10.50
8	$12.00
9	$13.50

A second way is to use an open sentence. Observing, again from the table, that one window is $1.50, the open sentence to solve this would be

$$9 \times 1.50 = \underline{\quad\quad}$$

and a simple calculation could give the answer of $13.50.

A third way would be to observe that three windows is $4.50, and six windows is $9.00. Since $3 + 6 = 9$, the amount Roger will charge for nine windows would be the sum of those two amounts:

$$\$4.50 + \$9.00 = \$13.50$$

2. June's gumball machine gives two gumballs for each penny. One way of solving this problem would be to write an open sentence:

$$2p = g$$

where p stands for pennies and g stands for gumballs. If she uses 4 pennies, then, she will get

$$2 \times 4 = 8 \text{ gumballs out of the machine}$$

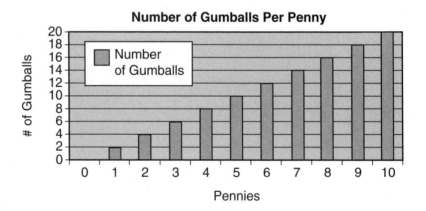

Number of Gumballs Per Penny

3. Frederica would find that the temperature is 95°F at 2 P.M. One way of solving this would be to use a table:

Hour	Temperature (°F)
9 A.M.	70
10 A.M.	75
11 A.M.	80
12 P.M.	85
1 P.M.	90
2 P.M.	95

Or a bar graph like the following:

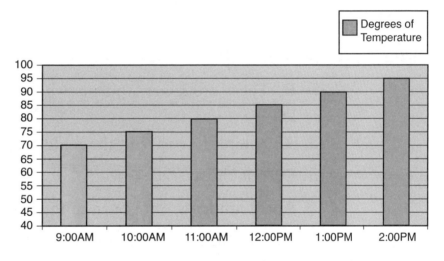

4. Jiminez would find that, at 6 P.M., the tank on his hay bailer would have 8 gallons. One way to solve this is with a table.

Hour	Gallons
2 P.M.	16
3 P.M.	14
4 P.M.	12
5 P.M.	10
6 P.M.	8

5. Carlita would find that the three missing numbers are 380, 320, and 260.

One way of finding the missing numbers would be to observe that the next number in the sequence goes down by 20, so the next number after 400 is **380**, and then 360, 340, **320**, 300, 280, **260**, and 240.

A second way of finding this is to use an open sentence. To find the next number in the sequence, the student would solve this open sentence:

$$a - 20 = b$$

where a is a known number in the sequence and b is the next number in the sequence.

CHAPTER 5

ANSWERS TO DATA ANALYSIS EXERCISES

1. **B.** There are fewer students who like mushrooms than plain.

2. **A.** A pink flower is in the middle.

3. **D.** 6 is the average (mean).

4. **A.** 12 students were born in the first half of the year.

5. **D.** The SUVs are the most popular vehicles on Gravel Rd.

ANSWERS TO PROBABILITY EXERCISES

1. **C.** 1 in 4. The probability is 13 in 52, which reduces to 1 in 4.

2. **D.** 1 in 3. The probability is 5 in 15, which reduces to 1 in 3.

3. **A.** 1 in 6. This is a straight probability.

4. **B.** 1 in 10. The probability is 2 in 20, which reduces to 1 in 10.

5. **D.** 1 in 2. The probability that Janice will get into an odd-numbered gondola is 6 in 12, which reduces to 1 in 2.

6. **A.** 1 in 13. The probability that she will draw a king from a standard deck of cards is 4 in 52, which reduces to 1 in 13.

ANSWERS TO OPEN-ENDED QUESTIONS: DATA ANALYSIS AND PROBABILITY

1. The line graph looks like this:

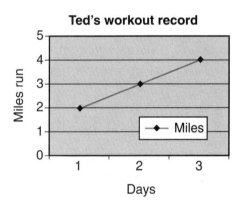

If he continues the pattern, he will run 5 miles on day 4 and 6 miles on day 5.

2. Half of the students took apple juice. Based on the graph, and knowing that there are 24 students in the class, there were $24 \times \frac{1}{2} = 12$. Twelve students took apple juice. Now, since the whole class is 100% of the class, one-half of the class is half of 100, or 50% of the class.

There were $\frac{1}{6}$ of the students who took orange juice. Based on the graph, and knowing there are 24 students in the class, there were $24 \times \frac{1}{6} = 4$. Four students took orange juice. Since the whole class is 100% of the class, one-sixth of the class is 16.7%.

There were $\frac{1}{3}$ of the students who took cranberry juice. Based on the graph, and knowing there are 24 students in the class, there were $24 \times \frac{1}{3} = 8$. Eight students took cranberry juice. Since the whole class is 100%, $\frac{1}{3}$ is 33.3% of the class.

3. A. To answer this question, we must add all 8 hours that are recorded:

$$10 + 25 + 55 + 55 + 45 + 60 + 55 + 50 = 355,$$
so 355 bags of nuts were sold that day

B. Rob sold the most amount at 3 P.M.: 60 bags. He sold 25 bags at 11 A.M., so $60 - 25 = 35$ bags. Rob sold 35 more bags of nuts at 3 P.M. than at 11 A.M.

C. There were three hours during which an equal number of bags were sold. The three hours are 12 P.M., 1 P.M., and 4 P.M. In each of those hours Rob sold 55 bags, so the total number of bags he sold in those three hours was $55 \times 3 = 165$.

4. A. If a contestant chooses 1 and 2, he has 2 in 8 chances of winning, and that reduces to 1 in 4 chances.

B. If a contestant places money on all the odd numbers, she will have 4 in 8 chances of winning, which reduces to 1 in 2 chances.

C. If a contestant chooses one number, he has 1 in 8 chances of winning.

ANSWERS TO DISCRETE MATHEMATICS COUNTING AND LISTING EXERCISES

1. **C.** There are six arrangements: M-1, D-2, B-3; M-1, D-3, B-2; M-2, D-1, B-3; M-2, D-3, B-1; M-3, D-1, B-2; M-3, D-2, B-1

2. **D.** Owens

3. **A.** 12 + 13 = 25, and 25 – 18 = 7. So, 7 students have both dictionaries and encyclopedias.

4. **B.** There are five students who like red, more than any other color.

5. **A.** P-P, P-N, N-N, P-D, D-D, N-D

6. **D.** 9 + 10 = 19, and 19 – 14 = 5. So, five pictures were both in brown frames, and were painted mostly with red paint.

ANSWERS TO DISCRETE MATHEMATICS: VERTEX-EDGE GRAPHS AND ALGORITHMS EXERCISES

1. **D.**

2. **B.**

3. **C.**

ANSWERS TO OPEN-ENDED QUESTIONS: DISCRETE MATHEMATICS SYSTEMATIC LISTING, COUNTING, VERTEX-EDGE GRAPHS, AND ALGORITHMS

1. Mac can come up with many combinations of nickels, dimes, and quarters to equal 90¢. If he uses only nickels, he'll use 18. If he uses only dimes, he'll use 9. The smallest number of coins he could use is three quarters, one dime, and one nickel.

$$25 + 25 + 25 + 10 + 5 = 90$$

Any combination of coins that will add up to 90 will be acceptable, but the equation must demonstrate that the value of the coins shown will add up to 90.

2. Samantha's class has 19 students. If 11 students have Dandycakes, and sixteen students have an orange, then $11 + 15 = 26$. Now, 26 is more than 19, and that means that some people have both. Subtract 19 from 26: $26 - 19 = 7$. This means that seven students have both a Dandycake and an orange. Showing this equation is important. Since seven students have both, the number of students who have only a Dandycake is $11 - 7 = 4$.

The number of students who had only an orange is $15 - 7 = 8$. The solution can also be shown as a Venn diagram.

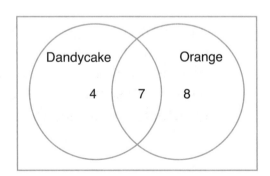

3. Alan has three jackets and five ties, so matching up each of the three jackets with each of the five ties will give us fifteen combinations, total.

Another way to solve this would be to multiply $3 \times 5 = 15$, although all the combinations should be shown.

4. There are ten ways to go from Arturo's Pizza Shop to the middle school.

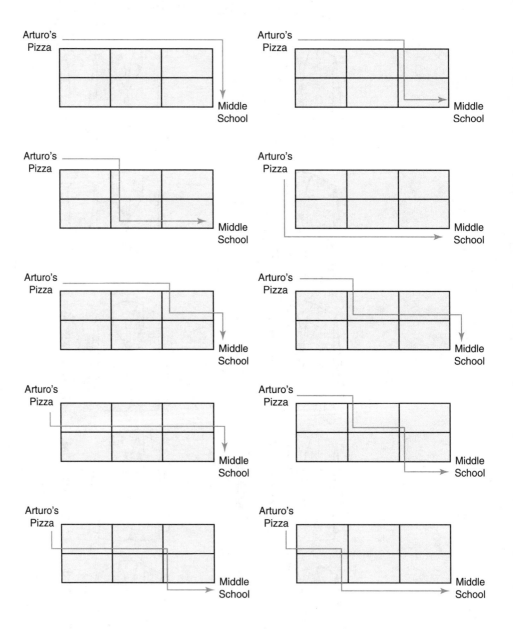

CHAPTER 6

ANSWERS TO PROBLEM-SOLVING EXERCISES

1. Greg is planning his barbeque. He needs to decide when to put the various items on his grill. It would seem that if we worked backward from 3:00 P.M., we'll get our answer.

Hamburgers	Ribs	Hot Dogs	Chicken
3:00 P.M. − 20 min 2:40 P.M.	3:00 P.M. − 35 min 2:25 P.M.	3:00 P.M. − 10 min 2:50 P.M.	3:00 P.M. − 30 min 2:30 P.M.

So, if he starts at 2:25 P.M. putting the ribs on, then at 2:30 P.M. he puts the chicken on, at 2:40 P.M. he puts the hamburgers on, and finally at 2:50 P.M. he puts the hot dogs on, he'll be able to take all of them off at 3:00 P.M.

2. Childrens' admission price is $7 and the adults' price is $12, so some combination of sevens and twelves that add up to 104 should give us our answer. One restriction we need to remember is that there were more children than there were adults. Guess and check appears to be a good way to solve this. Let's try seven children and six adults:

$$7 \times 7 = 49$$
$$12 \times 6 = \underline{60}$$
$$\$109$$

That's too much. But, we see that we'll get an odd number if we multiply 7 by an odd number. We need to multiply 7 by an even number to get 104. Let's try eight children and five adults:

$$7 \times 8 = 56$$
$$12 \times 5 = \underline{60}$$
$$\$116$$

Again, that's too much. But, it's too much by just $12. Let's drop one of the adults:

$$7 \times 8 = 56$$
$$12 \times 4 = \underline{48}$$
$$\$104$$

We've got it. Do you see how, in each case, the previous guess helped to narrow choices for the next guess, and helped us find the answer? Always try to think of how the last guess helps you toward the next guess. This is one of the most powerful aspects of the guess and check method: its ability to narrow our choices, getting us closer to the answer with each guess.

3. We need to find out how far they have kayaked and how far they are from their starting point. Drawing a picture would seem to work for this one, but not before we calculate how far they have paddled one way and the other.

They paddled eastbound from 8:00 A.M. till noon. That is 4 hours. They went 5 mph during that time. $4 \times 5 = 20$ mi. They traveled 20 miles east. Then, they traveled from 1:00 P.M. till 5:00 P.M. That is 4 hours. They traveled an a speed of 4 mph during that time. $4 \times 4 = 16$ mi. They traveled 16 miles west. So, let's draw a picture of this:

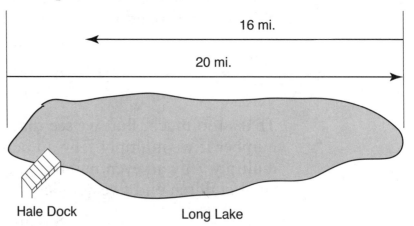

As you can see, Phil and Ben are not at their original starting place. To find out how short they are, subtract 16 from 20: 20 – 16 = 4. They are 4 miles short of their starting point. This is a good example of a problem that can be solved by a combination of a picture (or drawing) and a few simple equations.

4. Andrew picks up 6 bottles/cans for every 5 bottles/cans Henry picks up, and we know that the total picked up is 132. We want to know how many each boy picked up. If we made up a table, we could keep track of how much each boy picked up and of the running total.

Andrew	Henry	Total
6	5	11
12	10	22
18	15	33
24	20	44
30	25	55
36	30	66
42	35	77
48	40	88
54	45	99
60	50	110
66	55	121
72	60	132

So, Andrew picked up 72 bottles/cans, and Henry picked up 60 bottles/cans. This gives us a total of 132 bottles/cans.

5. We need to find out how many dots are in an array that has 10 on a side. A chart, along with a picture, would help us to see the solution to this.

As you can see, a triangle with one dot has one dot in it, an array with two dots on a side has three dots in it, and an array of dots having three dots on a side has 6 dots in it. Let's look at this in a table:

Triangle	Dots
1	1
2	3
3	6
4	10
5	15
6	21
7	28
8	36
9	45
10	55

So, the answer is that a triangular array of dots that has 10 dots on a side will contain 55 dots.

ANSWERS TO TEST YOUR SKILLS: PROBLEM SOLVING AND OTHER MATH SKILLS

For all of these problems we'll use the four-step process.

1. a) Make sense of the problem: We're trying to find the beginning number in a sequence of operations. What are the facts? We have the sequence of numbers, and the operations, but not the beginning number. We know all the operations and all the numbers, except the beginning one. How can we do this?

   b) Make a plan: Can we draw a picture? That won't help. Can we look for a pattern? Not really, because there is no pattern to see. Can we work backward? Well, that might help, but we need to know what we started with. Can we guess and check? Yes, that seems to be the right way to proceed.

   c) Carry out the plan: So, let's try a starting number. The number should be odd, because we add 5 right away, and then divide by 2, so we need an even number. Starting with an odd number will insure an even number when we have to divide by 2. Let's try 9:

   $9 + 5 = 14 \div 2 = 7 - 4 = 3 + 9 = 12 \times 2 = 24 \div 5 =$ Oh, oh. We can't divide 24 by 5 evenly, and our final answer is 8. We know that 9 does not work, but we need a number larger than that number. Let's try 15: $15 + 5 - 20 \div 2 = 10 - 4 = 6 + 9 = 15 \times 2 = 30 \div 5 = 6$. We're not far away from 8 (just two off). But, by using a multiple of 5, we seemed to get a whole number for an answer. Let's try 25: $25 + 5 = 30 \div 2 = 15 - 4 = 11 + 9 = 20 \times 2 = 40 \div 5 = 8!$ We've found it. Did you see that we gleaned clues as we went along for how to "choose smarter" on the next pick? Try to do that all the time.

   d) Check your answer: does this work? Yes, because when you plug the number in you'll get 8. Does this make sense? Yes, it does. Are there other numbers that would work? Probably not. So, this works.

2. a) Make sense of the problem: We're trying to find the perimeter to get the proper amount of fence. The garden is shaped like a trapezoid. How will we do this?

b) Make a plan: This is not a difficult problem. If we get the perimeter, we'll know the amount of fence we need. We'll add up all the legs of the trapezoid. It might be helpful to draw a diagram of the garden.

c) Carry out the plan: We have $4 + 6 + 5 + 8 = 23$. That's 23 feet. When you know units, you should put them in.

d) Check your answer: Does this make sense? Yes it's reasonable that 23 feet of fence will adequately fence in this garden. Are there other answers that will work? Not really, because we need an exact amount, here. So this works.

3. a) Make sense of the problem: We need to know how many numbers between 100 and 200 have a 6 in them. What are the facts? The numbers from 100–200. How will we do this?

b) Make a plan: Can we guess? That won't work, because we'll have no good way to check it, except to look at all the numbers, which will take a long time... Can we make a picture? A picture won't solve the problem. Can we look for a pattern? Yes, that probably will get the answer for us. So let's try a pattern.

c) Carry out the plan: How will we set up the plan? Let's look at the first ten numbers. 100, 101, 102, 103, 104, 105, 106, 107, 108, 109, 110. So, for the first ten numbers, a 6 appears once. So, for each group of ten numbers, a 6 will appear once. Is this true? Let's check the next ten numbers: 111, 112, 113, 114, 115, 116, 117, 118, 119, 120. Yes, that works. There are ten groups of these tens, so in the ten groups of tens from 100–200, there are ten numbers that have a 6 in them. Is that right? Oh, wait! When we get to the numbers

160 to 169, all ten numbers have a 6 in them. OK, so there are ten more numbers: $10 + 10 = 20$. Is that right? Well, not really, because one of those numbers, 166, is being counted twice, so we need to take one away from that 20. There are 19 numbers between 100 and 200 that have a 6 in them.

d) Check your answer: Does this make sense? Well, we looked at the numbers, found a pattern, and followed that pattern. It's reasonable that there are 19 numbers that have a 6 in them.

4. a) Make sense of the problem: What is the required answer? We want the last number in the sequence. What are the facts? We have the sequence of numbers. How can we get this? Figure out the pattern (how we get from one number to the next) and then apply that rule to the last number to get the required number.

b) Make a plan: How will we solve this? Searching for a pattern seems to be the best way to approach this problem. We'll examine each number, and see how to get from one number to the next, and see what the pattern is to do that. Then we'll try to apply that pattern to the next number in the sequence.

c) Carry out the plan: Let's look at all the numbers in the sequence:

$$23 \quad 28 \quad 26 \quad 31 \quad 29 \quad 34 . . .$$

This is a very strange sequence. It goes up, then down, then up again. What number does it go up by? 5, initially, then down 2, then up 5, then down 2, then up 5 again. So, the next number in the sequence would be down by 2 or 32.

d) Check your answer: Does this sound reasonable? Yes, it does. The pattern is: 5 up, 2 back, and repeat. So following that pattern, the next number will be the last number minus 2, or 32.

5. a) Make sense of the problem: What is the required answer? We want to find out the distance from each side of the front yard to plant a tree. What are the facts? We have the diameter of the dirt ball, and the width of the yard. How can we get this? We can draw a picture to find out where to plant the dirt ball of the tree.

b) Make a plan: We'll draw a picture so that we can figure how far from each side Peter needs to put the dirt ball so that it is in the center of the yard. Peter might also look at how to center it using only numbers.

c) Carry out the plan:

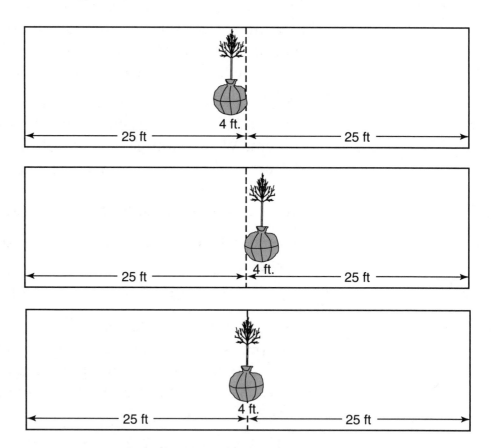

You see that if Peter puts the dirt ball at 25 ft., it would extend 4 ft. on one side or the other, so it wouldn't be centered. He must place it so that 2 ft.

are extended on each side of the center, so that it is centered.

There is another way. The yard is 50 ft. long, and half of that is 25. The dirt ball is 4 ft., so cut it in half, giving 2 ft. on each side. So the dirt ball will start at 23 ft. and end at 27 ft. on the number line, and that will center it.

d) Check your answer: Does this answer the question? Yes, it does. It places the dirt ball exactly in the center of the yard. Peter saw two ways of doing it here, and many problems can be done in more than one way.

6. a) Make sense of the problem: What is the question being asked? How many different sums of money can be made with four coins? What are the facts? Mickey has 2 pennies, a nickel, and a quarter. Note that using one or the other penny will not give another sum of money. Mickey wants only to find different sums.

b) Make a plan: Will a picture help? Not really. A chart probably will be best, because it will show all the possibilities very clearly. So Mickey will make a chart.

c) Carry out the plan: Here's the chart

1 cent	1	2	1	2	1	2		1	2
5 cents			1	1			1	1	1
25 cents					1	1	1	1	1
sum	1	2	6	7	26	27	30	31	32

So we have 9 sums.

d) Check your answer: Does this give all the sums? Yes, it does. Mickey looked at all the possible combinations from one of each coin to all four coins together. All the possible sums are here.

7. a) Make sense of the problem: What is the desired quantity? How many free tickets will be given? What are the facts? We know that for every 8 tickets, we get one free, and we need 50 tickets. How will we do this?

b) Make a plan: If we take groups of 8 away from the school group, and then take away one more (to account for the free ticket, 9 in all), we'll find out how many free tickets we will get.

c) Carry out the plan: We start with 50:

$$50 - 9 = 41$$
$$41 - 9 = 32$$
$$32 - 9 = 23$$
$$23 - 9 = 14$$
$$14 - 9 = 5$$

5 – 9 ? We cannot take 9 from 5, so we're done. We don't have another 8 students to get a free ticket, so we'll get 5 free tickets.

d) Check your answer: Does this answer the question? Yes, it does. See how repeated subtraction will get us to our answer.

8. a) Make sense of the problem: What are we to find? We need to name the shape formed by the four points, and we need to find the perimeter of the shape. Then, we need to slide the figure up four units, and calculate where the points would be that form its corners.

b) Make a plan: The best way to solve this is to plot the points on a graph and connect them. Then, we can see what shape is made.

c) Carry out the plan:

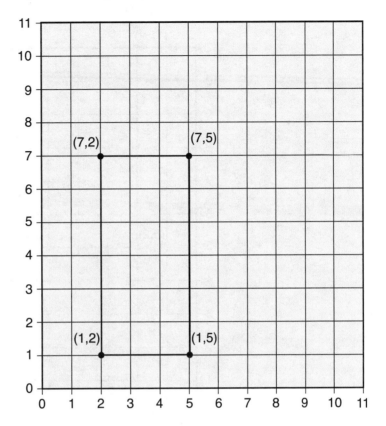

A. We see that the figure made is a rectangle.

B. We see that the rectangle's width is 3 units, and the length is 6 units, so the perimeter is $3 + 6 + 3 + 6 = 18$.

C. If we slide the rectangle up four units, we get this

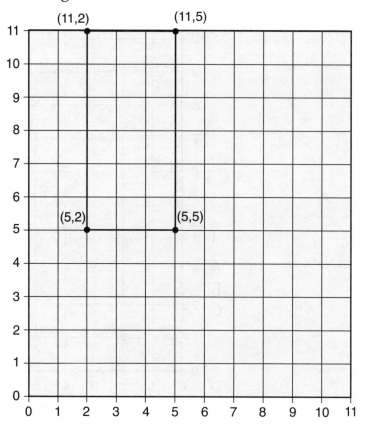

d) Check your answer: Does this answer the questions? Yes, it does. Does it seem reasonable? Yes.

9. a) Make sense of the problem: What is the required answer? It's to know how many books the bookstore sold on the first day, what day the sales went down, and how many books were sold for all six days we are looking at. What are the facts? We have a pictograph where each book represents 10 books. How will we do this?

b) Make a plan: It would seem that a chart, or table, will show the amounts sold more clearly. A chart is what we should use. We should expand the number of books out, however, giving the proper number of books for each.

c) Carry out the plan: Here is the chart:

Day	Books Sold
1	30
2	45
3	50
4	60
5	55
6	70

So, the answers to the three questions are:

    A. Bob and Ray sold 60 books on the fourth day.
    B. Sales went down on the fifth day.
    C. The total amount of books sold in the six days was:

$$30 + 45 + 50 + 60 + 55 + 70 = 310 \text{ books}$$

d) Check your answer: Does this answer all the questions? Yes, it does. See how easy it is when we put it in table form. It's much easier to read.

10. a) Make sense of the problem: What is the required answer? How many different combinations of outfits can Martina make? What are the facts? The number of blouses Martina has (4), and the number of pants she has (3). How will we do this?

b) Make a plan: The way we can solve this is to make a tree diagram.

c) Carry out the plan: Martina will make a diagram that matches up all the blouses with all the pants.

You see that there are 12 combinations of blouses with pants.

If she had two pairs of shoes, how many combinations of blouses, pants, and shoes will Martina have?

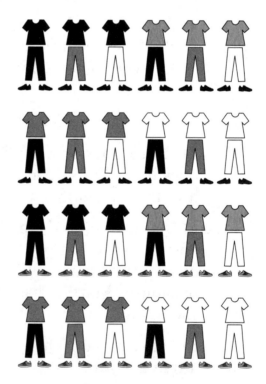

You see that there are 24 combinations of blouses, pants, and shoes.

d) Check your answer: Does this make sense? Yes, it does. You can see clearly, with a tree diagram, *all* the combinations that are available to Martina when she goes on vacation.

Chapter

8

SAMPLE TESTS

1

SAMPLE TEST #1

SESSION 1—DAY 1

DIRECTIONS:

Choose the best answer of the answer choices given for each of the following problems. Fill in the circle next to your choice. You may NOT use a calculator.

1. In the number 319,042, which digit is located in the hundreds position?

   Ⓐ 0

   Ⓑ 3

   Ⓒ 2

   Ⓓ 1

2. In the number 790.26, which digit is located in the hundredths position?

   Ⓐ 7

   Ⓑ 0

   Ⓒ 6

   Ⓓ 9

Go On

3. Which number would be correct if we were rounding $329.49 to the nearest dollar?

Ⓐ $330.00

Ⓑ $329.00

Ⓒ $300.00

Ⓓ $400.00

4. Lauren went to the Joe's Pizzeria with two of her friends. They decided to purchase a large eight-slice pizza with pepperoni and mushrooms on top. When Lauren and her friends finished eating, they still had two slices left over. Which fraction below represents the amount of pizza that Lauren and her friends ATE?

Ⓐ $\dfrac{2}{8}$

Ⓑ $\dfrac{1}{4}$

Ⓒ $\dfrac{6}{8}$

Ⓓ $\dfrac{8}{8}$

1

5. Which number listed below is even?

Ⓐ 413,099

Ⓑ 307,413

Ⓒ 9,370

Ⓓ 71,211

6. Which of the following groups of numbers is arranged in order from LEAST to GREATEST?

Ⓐ 5,901, 4,333, 6,981, 8,328

Ⓑ 2,542, 4,561, 8,527, 8,519

Ⓒ 13,463, 17,322, 23,001, 27,443

Ⓓ 9,871, 6,420, 5,616, 11,742

Go On

7. Joseph and Kyle were comparing the number of MP3 songs they had downloaded onto their computers. After going through his music folder on his computer, Joseph counted a total of 317 songs. Kyle did the same thing on his computer and realized he had 511 songs. How many more songs does Kyle have than Joseph?

Ⓐ 194

Ⓑ 206

Ⓒ 828

Ⓓ 204

8. Teresa went to the cell phone store with her parents to purchase cell phones. Her parents planned on purchasing three phones: one for Teresa and one for each of them. Teresa's mother picked out a red phone that cost $319.00. Her dad picked out a silver one for $299.00. Teresa picked out a nice pink phone, which cost $149.00. Given the three prices of these phones, how much is it going to cost Teresa's parents in all?

Ⓐ $776.00

Ⓑ $676.00

Ⓒ $675.00

Ⓓ $767.00

If you have time, you may review your work in this section only.

SESSION 2–DAY 1

DIRECTIONS:

Choose the best answer of the answer choices given for each of the following problems. Fill in the circle next to your choice. You may NOT use a calculator.

9. Joel decided to make a summer job of mowing lawns for his neighbors. He started with just one lawn and was making $35 per week for cutting it. The other neighbors were so impressed with the great job he did that they too decided to hire Joel to cut their grass. Now Joel had a total of seven neighbors who were paying him weekly to cut their grass. How much money was Joel making every week for cutting those lawns?

Ⓐ $42.00

Ⓑ $28.00

Ⓒ $215.00

Ⓓ $245.00

2

10. If a jet ski can hold a maximum of two people at a time, how many jet skis would you need if there were 15 people who all wanted to ride at the same time?

Ⓐ 7

Ⓑ 8

Ⓒ 5

Ⓓ 15

Go On

11. Which of the following is a hexagon?

Ⓐ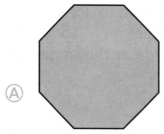

Ⓑ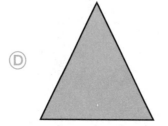

Ⓒ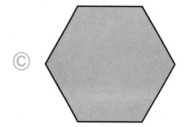

Ⓓ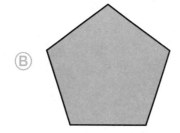

12. Which of the following figures represents only a rotation?

Ⓐ    Ⓒ

Ⓑ    Ⓓ

Go On

13. A normal-sized paper clip would measure about how much?

   Ⓐ 5 centimeters

   Ⓑ 5 millimeters

   Ⓒ 5 decimeters

   Ⓓ 5 kilometers

14. What unit of length would you use if you wanted to measure the perimeter of your classroom?

   Ⓐ Kilometers

   Ⓑ Meters

   Ⓒ Decimeters

   Ⓓ Millimeters

If you have time, you may review your work in this section only.

DIRECTIONS:

You are allowed to use your calculator for the following multiple-choice question in this part only.

15. Dylan wanted to build a dog pen for his new puppy. With time, the puppy would grow into a fairly large dog so he wanted to make sure to build a pen that would fit his new dog for years to come. After lots of thought, Dylan decided to make the dog pen in the shape of a rectangle, 15 ft × 25 ft. All he needed now was some fencing to put around it. How many feet of fence would Dylan need to cover the perimeter of the dog pen?

Ⓐ 60 ft.

Ⓑ 40 ft.

Ⓒ 10 ft.

Ⓓ 80 ft.

If you have time, you may review your work in this section only.

1

SESSION 1–DAY 2

DIRECTIONS

Choose the best answer of the answer choices given for each of the following problems. Fill in the circle next to your choice. You may NOT use a calculator.

16. Ramon wanted to install a new brick patio behind his house. He decided the patio would be 18 ft. × 32 ft. The store where he is buying the bricks sells them by the square ft. How many square ft. of bricks does Ramon need to complete his new patio?

Ⓐ 576 sq. ft.

Ⓑ 50 sq. ft.

Ⓒ 566 sq. ft.

Ⓓ 80 sq. ft.

Go On

17. Frank's parents called him from Florida. During the conversation, they told Frank that the temperature there in Florida was 96 degrees. Frank looked at his thermometer in his New Jersey house and saw that it was 77 degrees at his house. How much hotter was it in Florida than in New Jersey?

  Ⓐ 21 degrees

  Ⓑ 16 degrees

  Ⓒ 19 degrees

  Ⓓ 29 degrees

18.

What would you expect the next three shapes/colors to be if the pattern continued as shown? Explain your answer.

_____

_____

_____

19. What are the next three numbers in the pattern: 874, 884, 894, . . . ?

_____

_____

_____

If you have time, you may review
your work in this section only.

SESSION 2–DAY 2

DIRECTIONS:

You are allowed to use your calculator for the following open-ended questions in this part only.

20. Samantha's Superstore is having a storewide sale today only. Everything is marked down the same amount. Kelly spotted a pair of shoes that were originally $90 and are now on sale for $45. She also spotted a new TV set that was originally $300 and now is marked down to $150. Her sister is buying a DVD player that was originally $110, now on sale for $55. She was also interested in purchasing a new cordless phone that was marked as originally $60, but there was no sale sticker on it. If the discount for the phone is the same as the discount for all of the other items, what would the sale price of the phone be? Explain your answer.

2

21. Robby wanted to fill up his new 4-foot-deep swimming pool. The pool guy told him the water would fill at a rate of 6 inches per hour. How many hours would it take Robby to fill up his pool with water?

If you have time, you may review your work in this section only.

DIRECTIONS:

Choose the best answer of the answer choices given for each of the following problems. Fill in the circle next to your choice. You may NOT use a calculator.

22. If a function machine stated that the rule was +4, and it gave you the number 41 as the "OUT" number, what would the "IN" number have been?

Ⓐ 45

Ⓑ 35

Ⓒ 37

Ⓓ 39

23. Matthew, Sarah, Jamaal, and Elizabeth all get a weekly allowance from their parents. Matthew gets $5 a week, Sarah gets $5 a week, Jamaal gets $10 a week, and Elizabeth gets $20 a week. What is the mean (average) amount of allowance per week?

Ⓐ $5

Ⓑ $10

Ⓒ $15

Ⓓ $20

24.

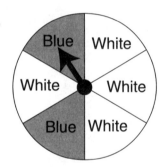

In the figure above, what is the probability that the spinner will land in the blue section?

Ⓐ $\frac{6}{1}$

Ⓑ $\frac{2}{6}$

Ⓒ $\frac{1}{6}$

Ⓓ $\frac{6}{6}$

Go On

25. A deck of playing cards has four Aces in it. Jerry's dad decided to play a game of "52-pickup" with him. Jerry didn't know what type of game that was but his father said he would show him. Jerry's father picked up a deck of cards and flung them all on the ground! He said "go ahead and pick them up!" Realizing now that "52-pickup" means you are picking up 52 cards, Jerry decided to go ahead and pick up the cards. What is the probability that Jerry will pick up an Ace as his first card?

Ⓐ $\dfrac{1}{52}$

Ⓑ $\dfrac{52}{4}$

Ⓒ $\dfrac{13}{1}$

Ⓓ $\dfrac{4}{52}$

26. What is the least number of coins you could use to make $0.42?

    Ⓐ 5

    Ⓑ 6

    Ⓒ 4

    Ⓓ 9

27. What does the r equal in the equation $r - 7 = 6$?

    Ⓐ 1

    Ⓑ −1

    Ⓒ 13

    Ⓓ 12

Go On

28. Stacy and Jim decided to each run a lemonade stand on their street. Stacy was going to sell her lemonade for $0.50 per glass. Jim thought he could sell more lemonade if he sold his lemonade for less, so he sold his for $0.25. Every time they sold a glass of lemonade they would make a tally to keep track of it. At the end of the day, Stacy had sold 23 glasses and Jim sold 39 glasses. Who made more money? Explain your answer.

29. If you had a standard American quarter and you flipped it in the air four times in a row, and all four times it landed on heads, what would be the probability that it would land on heads again the next time you flipped it? Explain your answer.

If you have time, you may review your work in this section only.

EXPLANATORY ANSWERS TO SAMPLE TEST #1

1. **A.** The 0 is in the hundreds position in the number 319,042.

2. **C.** The 6 is in the hundredths position in the number 790.26.

3. **B.** To determine whether we round up or down, we must look at the digit to the right of the ones position. In this case it is a 4. 4 is below 5 (half) so we round down and the 9 remains the same, leaving us with $329.00. Another way to look at this is that $329.49 is closer to $329.00 than it is to $330.00.

4. **C.** Originally the pizza had 8 slices in it. If there were 2 slices left when they were finished eating, that means they ate 6 slices (8 – 2). So the portion of the pizza they ate is $\frac{6}{8}$, or 6 out of 8 slices.

5. **C.** To determine if a number is even, we look at the digit furthest to the right. If that digit is even, then the number itself is even. In this case the digit furthest to the right is a 0, which is even. Therefore, 9,370 is even.

6. **C.** The question asks us to select which answer has numbers that are arranged from LEAST to GREATEST. This is the only choice where the numbers increase. 27,443 is bigger than 23,001, which is bigger than 17,322, which is bigger than 13,463.

7. **A.** If Kyle has 511 songs and Joseph has 317 songs, we must subtract 317 from 511 to determine how many more Kyle has than Joseph. 511 – 317 is 194. So Kyle has 194 more songs than Joseph.

8.   **D.** In this problem Teresa's parents are buying three cell phones. The prices of the phones are listed at $319.00, $299.00, and $149.00. To find the total, we must add those numbers together to get the final answer of $767.00.

9.   **D.** Joel is making $35 per lawn, and he is mowing a total of seven lawns. To find the grand total, we must multiply 35 × 7, which gives us a final answer of $245.00.

10. **B.** If a jet ski can hold 2 people at a time, and 15 people want to ride at the same time we must divide it out to figure out how many jet skis we will need. 2 goes into 15 7 times (which gives us 14, meaning 14 people would be able to ride on jet skis). This isn't enough though because we have 15 people. So besides the 7 jet skis we will need for 14 people, we will also need one more jet ski for the last person who will be riding alone. So the grand total comes out to 8 jet skis.

11. **C.** A hexagon is a polygon made up of six sides (*hex* means 6).

12. **C.** This is the only choice where an object (the key) is rotated.

13. **A.** 5 centimeters (cm).

14. **B.** Kilometers may be used for measuring a long distance, such as driving. A decimeter may be used for measuring objects such as a window. A millimeter is used for measuring very small objects such as an eraser. None of those choices would be useful or convenient for measuring a classroom. Therefore, the correct answer is meters (remember, a meter is 39 inches, or roughly 3 feet).

15. **D.** A rectangle 15 ft. × 25 ft. has a total of four sides. The two shorter sides are 15 ft., and the two longer sides are 25 ft. To find the perimeter (distance around) we must add up all of the sides. 15 + 15 + 25 + 25 = 80. So the total amount of fence needed for the dog pen is 80 ft.

16. **A.** If the patio is shaped like a rectangle and measures 18 ft. × 32 ft., to find the area we would need to multiply these two numbers (area = length × width). 18 × 32 = 576. So Ramon would need 572 sq. ft. of bricks to build his new patio.

17. **C.** If the temperature in Florida is 96 degrees and the temperature in New Jersey is 77 degrees, we must subtract the two to find out the difference. 96 – 77 is 19. So it is 19 degrees hotter in Florida than it is in New Jersey.

18. The pattern is: blue circle, gray rhombus, white circle, black circle, blue triangle. This pattern begins to start over next when it lists the blue circle again. So the next four shapes/colors would be a continuation of the pattern: gray rhombus, white circle, black circle, blue triangle.

19. To figure out the pattern with the numbers 874, 884, 894, we must compare the numbers side by side. 884 is 10 more than 874 (rule is +10 so far). As long as the rule pans out again with the next number, we can confirm it to be true. 894 is 10 more than 884, so the rule in this pattern is +10. To find out the next three numbers, we just continue by adding 10 to 894, which gives us 904, then 914, then 924. So the next three numbers in this pattern are 904, 914, 924.

20. From comparing the sale prices with the original prices, we can determine that the sale is $\frac{1}{2}$ off (or 50% off). The shoes were originally $90 and are now $45 (half off). The TV was originally $300 and is now $150 ($\frac{1}{2}$ off), and the DVD player was originally $110 and is now $55 (also $\frac{1}{2}$ off). So a sale of $\frac{1}{2}$ off would turn a $60 phone into a $30 phone. Kelly's sister can expect to pay $30 for the new phone.

21. Robby's pool is 4 ft. deep. 4 ft. is equal to 48 inches. We are converting it to inches so we can compare it with the fill rate which was given in inches as well (6 inches per hour). So if the water fills at a rate of 6 inches per hour, we need to figure out how many 6's there are in 48. This will tell us how many hours it will take altogether. To figure this out, we merely divide 48 by 6 to get an answer of 8. There are eight 6's in 48 so it will take 8 hours to fill the pool up.

22. **C.** If the out number is 41 and we know the rule was +4, to figure out the answer we need to determine what number was added to 4 in order to get 41. To get the answer we can subtract 4 from 41 to get 37. 37 plus 4 (our rule +4) does equal 41.

23. **B.** To calculate the mean (average), we must add all of the numbers up then divide by the number of numbers we added. 5 + 5 + 10 + 20 is 40. We added up four numbers to get that answer so we must divide it by 4. 40 divided by 4 is 10, so the mean (average) of their allowances is $10.

24. **B.** The spinner has a total of six equal spaces on it, so 6 is the denominator (bottom number of the fraction). The top number is always what is being asked, in this case it was the probability of the spinner landing on blue (so the numerator is the number of spaces that are blue). There are two blue spaces, so the numerator is 2. So the probability of landing on blue is $\frac{2}{6}$, or two blue spaces out of a total of six spaces.

25. **D.** A full deck of cards has 52 cards in it (as stated in the problem). Four of those cards are Aces. So the probability (or chance) that Jerry will pick up an Ace on his first card is $\frac{4}{52}$. This means 4 Aces out of a total of 52 cards. So it is not overly likely that he will turn an Ace on the first card.

26. **A.** The least number of coins to make $0.42 is five. Those coins would be one quarter, one dime, one nickel, and two pennies.

27. **C.** $13 - 7 = 6$. In this type of algebraic equation and because addition is the opposite of subtraction, you can merely read it backwards as what does 6 plus 7 equal?

28. Stacy sold 23 glasses, and Jim sold 39. This does not mean that Jim made more money. Stacy sold her lemonade for $0.50 per glass, and Jim sold his for $0.25 per glass. To figure out what each one's grand total is, we must multiply. $23 \times \$0.50$ is $11.50. We can also look at this as "what is half of 23" since $0.50 is half of a dollar. For Jim we must multiply $39 \times \$0.25$ which is $9.75. Similarly, we can look at this as what is $\frac{1}{4}$ of 39, since $0.25 is $\frac{1}{4}$ of a dollar.

So Stacy's total is $11.50 and Jim's total is $9.75; therefore, Stacy made more money.

29. The probability that the quarter would land on heads is always going to be $\frac{1}{2}$. It does not matter what the flip history of that quarter is. Probability always remains the same, and for a quarter, the probability of it landing on either side will always be $\frac{1}{2}$. If for some reason it landed on heads nine times in a row, it still has a $\frac{1}{2}$ chance of landing on heads again— just as likely as it would be to land on tails.

SAMPLE TEST #2

SESSION 1–DAY 1

DIRECTIONS:

Choose the best answer of the answer choices given for each of the following problems. Fill in the circle next to your choice. You may NOT use a calculator.

---

1. In the number 91,023, which digit is located in the thousands position?

    Ⓐ 9

    Ⓑ 1

    Ⓒ 2

    Ⓓ 3

2. In the number 814.20, which digit is located in the hundredths position?

    Ⓐ 8

    Ⓑ 1

    Ⓒ 2

    Ⓓ 0

Go On

3. Which number would be correct if we were rounding $9,057.14 to the nearest hundred dollars?

    Ⓐ $9,100.00

    Ⓑ $9,000.00

    Ⓒ $9,200.00

    Ⓓ $9,060.00

4. Sammy purchased a bag of Skittles.® When he got home he decided to open the bag and sort them out by color. There were red, green, yellow, orange, and purple Skittles in the bag. He found that there were 10 Skittles of each color in the bag. What fraction of the entire bag were the red ones?

    Ⓐ $\dfrac{5}{1}$

    Ⓑ $\dfrac{1}{5}$

    Ⓒ $\dfrac{3}{5}$

    Ⓓ $\dfrac{2}{5}$

5. Which number listed below is odd?

   Ⓐ 6,444

   Ⓑ 140

   Ⓒ 730,412

   Ⓓ 89,201

6. Which of the following groups of numbers is arranged in order from LEAST to GREATEST?

   Ⓐ 7,210, 6,323, 6,232, 6,229

   Ⓑ 4,311, 2,690, 7,439, 5,678

   Ⓒ 107, 3,412, 3,421, 6,501

   Ⓓ 2,001, 2,010, 2,000, 2,613

If you have time, you may review your work in this section only.

SESSION 2–DAY 1

DIRECTIONS:

You may use a calculator for the following multiple-choice questions in this part only.

7. Jackie and her 3-year-old son Chase were going to Wally's Wacky Water Slides for some afternoon fun. The admission price was $17.95 for adults and $11.95 for children under the age of 5. How much would it cost for both of them to get into the Water Park?

    Ⓐ $18.90

    Ⓑ $29.80

    Ⓒ $19.90

    Ⓓ $29.90

8. Frank, John, and Maria went to the diner to get some appetizers. They ordered french fries for $6.95, mozzarella sticks for $8.95, and buffalo wings for $13.95. They also each got a soda, which was $2.00 each. How much was their bill that evening?

    Ⓐ $35.85

    Ⓑ $38.58

    Ⓒ $28.58

    Ⓓ $36.95

9. Paul opened up a power washing business this summer. With the business he would clean houses, decks, patios, and driveways using a power washer. If he charged $215.00 to power wash any house, how much would he make if he power washed 3 homes?

    Ⓐ $415.00

    Ⓑ $645.00

    Ⓒ $615.00

    Ⓓ $215.00

Go On

10. If a bus holds 52 people, what is the least number of buses that would be required in order for 113 people to be transported to a Broadway play?

Ⓐ 1

Ⓑ 2

Ⓒ 3

Ⓓ 4

If you have time, you may review your work in this section only.

SESSION 3—DAY 1

DIRECTIONS:

Choose the best answer of the answer choices given for each of the following problems. Fill in the circle next to your choice. You may NOT use a calculator.

11. Which of the following is a pentagon?

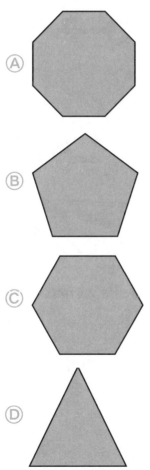

Go On

12. Which of the following figures represents only a flip?

13. What would be the best unit of length if you wanted to measure the thickness of your social studies textbook?

(A) Kilograms

(B) Kilometers

(C) Centimeters

(D) Grams

14. What unit of length would be the best choice if you wanted to measure the distance from your classroom to the principal's office?

(A) Meters

(B) Centimeters

(C) Kilograms

(D) Kilometers

Go On

15. Anthony was installing a new driveway at his house. The concrete company wanted to know how big the driveway would be so they could estimate how much it was going to cost. Anthony measured his driveway and came up with 22 feet wide by 36 feet long. How many square feet is the driveway?

Ⓐ 690 sq. ft.

Ⓑ 790 sq. ft.

Ⓒ 792 sq. ft.

Ⓓ 892 sq. ft.

If you have time, you may review your work in this section only.

DIRECTIONS:

You may use a calculator for the following multiple-choice questions in this part only.

16. Poppy decided she was going to drive to Maine to cash in on the $0.05 deposit deal for all of her soda bottles. In Maine you can turn in certain containers to the recycling center and they will pay you $0.05 for each one. Poppy had collected bottles for almost 3 months. She counted them all before heading to the recycling center and found she had a total of 96 bottles. How much money should she receive if she turns in the 96 eligible bottles and gets cash refund of $0.05 for each?

Ⓐ $0.05

Ⓑ $5.80

Ⓒ $6.00

Ⓓ $4.80

Go On

17. In May the water temperature at the Longport beach in New Jersey measured 54 degrees. In July that same beach's water temperature measured 77 degrees. How much warmer was the water in July than in May?

Ⓐ 23 degrees

Ⓑ 32 degrees

Ⓒ 131 degrees

Ⓓ 13 degrees

If you have time, you may review your work in this section only.

DIRECTIONS

For the following open-ended questions, write your answers in the spaces provided. Show your work. For the multiple-choice questions, choose the best answer of the answer choices given. Fill in the circle next to your choice. You may NOT use a calculator in this part.

18.

What would you expect the next three shapes/colors to be if the pattern continued as shown? Explain your answer.

_____

_____

_____

Go On

19. What are the next three numbers in the pattern: 7,511, 7,531, 7,551, . . . ?

_____

20. Angela went to Tim's Tire Shop to get new tires put on her car. She had very expensive 22-inch tires on her car that were $295.00 each. Luckily Tim's shop was running a deal where you could buy three tires and get one free. If Angela needed to replace all four tires on her car, how much would it cost her for the tires?

_____

21. If gasoline costs $4.00 per gallon, how much would it cost to fill up a 4×4 truck with a 36-gallon tank?

_____

22. If a function machine stated that the rule was × 6, and it gave you the "42" as the "OUT" number, what would the "IN" number have been?

Ⓐ 42

Ⓑ 4

Ⓒ 6

Ⓓ 7

23. If Isabel gets a $5 per week allowance, how much money would she have collected in a 7-week period?

Ⓐ $25

Ⓑ $35

Ⓒ $15

Ⓓ $20

If you have time, you may review your work in this section only.

SESSION 3—DAY 2

DIRECTIONS:

For the following open-ended questions, write your answers in the spaces provided. Show your work. For the multiple-choice questions, choose the best answer of the answer choices given. Fill in the circle next to your choice. You may NOT use a calculator in this part.

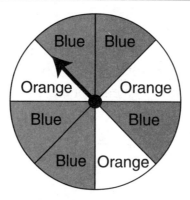

24. In the figure above, what is the probability the spinner will land in the orange section?

Ⓐ $\dfrac{1}{8}$

Ⓑ $\dfrac{8}{3}$

Ⓒ $\dfrac{3}{8}$

Ⓓ $\dfrac{3}{5}$

25. If a bag of marbles had 7 blue, 8 clear, 2 red, 1 white, and 7 orange marbles in it, what would be the probability that you would reach your hand into the bag without looking and pull out a red marble?

   (A) $\dfrac{1}{23}$

   (B) $\dfrac{7}{25}$

   (C) $\dfrac{2}{23}$

   (D) $\dfrac{2}{25}$

26. What is the least number of coins you could use to make $0.07?

   (A) 1 coin

   (B) 2 coins

   (C) 3 coins

   (D) 4 coins

Go On

27. At the Yogi Berra baseball field, to hit a homerun you have to reach the outfield fence, which is 110 feet from home plate. When Colin was at bat, he hit a homerun that went 17 feet past the outfield fence. How far did Colin hit the ball? Explain your answer.

_____

28. Outside of the local electronics store, 44 people are lined up to buy the new video game system, which has just been released. They have been camping outside of the store for over 24 hours. Just before the store is about to open, an employee comes outside and tells everyone they have 75 of the game systems in stock and there is a limit of one purchase per customer. If you are the 45th person in line, what is the probability that you will be able to purchase the game system when your turn comes up? Explain your answer.

_____

If you have time, you may review your work in this section only.

EXPLANATORY ANSWERS TO SAMPLE TEST #2

1. **B.** The 1 is in the thousands position in the number 91,023.

2. **D.** The 0 is in the hundredths position in the number 814.20.

3. **A.** $9,100.00 To determine whether we round up or down, we must look at the digit to the right of the hundreds position. In this case it is a 5, then a 7. 5 is the halfway mark and 7 is above 5 (half) so we round up, leaving us with $9,100.00. Another way to look at this is that $9,057.14 is closer to $9,100.00 than it is to $9,000.00.

4. **B.** There are two ways to look at this problem. There are 10 of each color, and 5 different colors. So the grand total amount in the bag is 50. The question is asking what fraction of the bag is red. Well, if 10 are red, and there are 50 altogether, then the fraction is $\frac{10}{50}$ which is simplified as $\frac{1}{5}$ (10 goes into 10 once, and 10 goes into 50 five times). The second way to look at this problem is comparing the colors together. There are 10 of every color in the bag, so their portions are equivalent to each other (the same number of each color). Because of this, we can see the answer as $\frac{1}{5}$ because red is one color out of the five colors in the bag. Note that this ONLY works because each color is represented with the same amount of Skittles (10). If a color were to have more or less than 10, this way would not work. They must be equal.

5. **D.** To determine if a number is odd, we look at the digit furthest to the right. If that digit is odd, then the number itself is odd. In this case the digit furthest to the right is a 1, which is odd. Therefore, 89,201 is odd.

6. **C.** The question asks us to select which answer has numbers that are arranged from LEAST to GREATEST. This is the only choice where the numbers increase. 6,501 is larger than 3,431, which is larger than 3,412, which is then larger than 107.

7. **D.** The admission price is $17.95 for adults and $11.95 for children under the age of 5. Jackie is an adult, and her son Chase is a child since he is under the age of 5. So we must add $17.95 (Jackie's ticket) and $11.95 (Chase's ticket) together to get a final answer of $29.90 for the two of their admissions.

8. **A.** We need to find the sum of their purchases at the diner. French fries are $6.95, mozzarella sticks are $8.95, and buffalo wings are $13.95. If they each got a soda, which is $2.00 a piece, then the total for sodas is $6.00 (3 of them times $2.00 per soda). So now we must add their totals. $6.95 + $8.95 + $13.95 + $6.00 for a total of $35.85.

9. **B.** If the price per house to power wash is $215.00, and 3 homes are power washed, we would multiply to find our final answer. $215 per house times 3 equals $645.00. You could also add $215 + $215 + $215 which is the same as multiplying $215 × 3.

10. **C.** If there are 113 people and each bus can hold 52, we need to figure out how many buses are required to transport everyone. Two buses would be able to hold 104 people (52 + 52). That's not enough because we have 113 people to transport. Three buses would be able to hold 156 people. This is our

answer. Since three buses can hold 156 people, and we have 113 to transport, this is the least number of buses possible to transport everyone.

11. **B.** A pentagon is a polygon made up of 5 sides (*pent* means "five").

12. **A.** This is the only choice where an object is flipped (the key).

13. **C.** Centimeters are the only unit of length given that would make sense if you were measuring the thickness of a textbook. Kilometers are for measuring very long distances. Kilograms and grams are used for measuring weight, not length.

14. **A.** Kilometers may be used for measuring a long distance, such as driving. A kilogram is used for measuring weight, not distance, and a centimeter is used for measuring a very small distance. Therefore, the correct answer is meters (remember, a meter is 39 inches or roughly 3 feet).

15. **C.** The cement company is trying to figure out how large the driveway is so they can determine how much concrete they will need to deliver. To do this, they had Anthony measure the length and width of the driveway. To figure out the area (or amount of space it takes up) you would use the formula Area = Length × Width. In this case, the driveway is 22 ft. × 36 ft. So the area would be 22×36 or 792 sq. ft.

16. **D.** If the recycling center in Maine is giving out $0.05 per bottle to help with recycling, and Poppy has 96 bottles, we would multiply $96 \times \$0.05$ to find out the value of all of her bottles. 96×0.05 is $4.80.

17. **A.** If the water temperature in May is 54 degrees, and it is 77 degrees in July, we must subtract the two to find the difference. 77 – 54 is 23. This means that the water is 23 degrees warmer in July than it is in May.

18. The pattern is: blue circle, white circle, blue triangle, then gray rhombus. This pattern begins to start over next when it lists the blue circle again. So the next 3 shapes/colors would be a continuation of the pattern: white circle, blue triangle, then gray rhombus.

19. To figure out the pattern with the numbers 7,511, 7,531, 7,551, we must compare the numbers side by side. 7,531 is 20 more than 7,511. So the pattern could possibly be +20. To be certain of this, we must test it on another number. 7,551 is 20 more than 7,531, so again the pattern +20 holds true. To find the next three numbers we merely add 20 consecutively to the end number (7,551). So the next three numbers in this pattern are 7,571, 7,591, and 7,611. You may want to do that last one on a scrap piece of paper since you will be carrying a 1 into the hundreds position to solve it.

20. Angela needs to purchase four tires for her car. Each tire is $295.00. Luckily the tire shop is running a deal where you can buy three and get one free. So for Angela to get four tires she needs to pay for three of them. If they are $295.00 each, we need to multiply $295.00 × 3, which gives us a total of $885.00. $885.00 is the cost of three tires, but remember that the tire shop is throwing that fourth tire in for free. So $885.00 is the final cost to put new tires on Angela's car.

21. If a 4 × 4 truck has a 36-gallon gas tank that means it holds 36 gallons of gas at one time. If gas is $4.00 per gallon and the truck needs 36 gallons, we must multiply to find out the total cost. 36 gallons × $4.00 per gallon = $144.00. So the total cost to put 36 gallons of gas into the truck is $144.00.

22. **D.** If the out number is 42 and we know the rule was × 6, to figure out the answer we need to determine what number was multiplied by 6 to get 42. If you know your multiplication facts, you will realize that 6 × 7 is 42. So the "IN" number had to be 7.

23. **B.** Isabel receives $5.00 per week. To find out how much money she would have after 7 weeks we need to multiply. $5.00 per week × 7 weeks is $35.00. So after 7 weeks, Isabel has received a total of $35.00 in allowance.

24. **C.** The spinner has a total of 8 equal spaces on it, so 8 is the denominator (bottom number of the fraction). The top number is always what is being asked, in this case it was the probability of the spinner landing on orange (so the numerator is the number of spaces that are orange). There are 3 orange spaces, so the numerator is 3. So the probability of landing on orange is $\frac{3}{8}$, or 3 orange spaces out of a total of 8 spaces.

25. **D.** If the bag of marbles has 7 blue, 8 clear, 2 red, 1 white, and 7 orange marbles in it, that is a total of 25 marbles (7 + 8 + 2 + 1 + 7). To figure out the probability of pulling out a red marble, we need to know how many marbles there are altogether and which color we are calculating the probability for. Since there are 25 marbles altogether, and there are 2 red marbles, the probability of pulling out a red marble is $\frac{2}{25}$ (or two chances in twenty-five).

26. **C.** The smallest number of coins we can make $0.07 with is 3: one nickel and two pennies.

27. To figure out the distance of Colin's hit, we must look at the information the problem gives us. The problem says that the fence is 110 feet from home plate. If Colin hits the ball 17 feet past the fence, then the total distance of his hit would be 110 ft. (to the fence) + 17 ft. (past the fence) for a total of 127 ft. So Colin hit the ball 127 ft.

28. If the store employee says they have 75 units in stock, and you are the 45th person in line, then the probability of you being able to purchase the unit when your turn comes up is 100%. The unit will still be in stock by the time you reach the door. In fact, there will be 30 left when you get to the checkout counter (45 people ahead of you subtracted from 75 units is 30).

INDEX

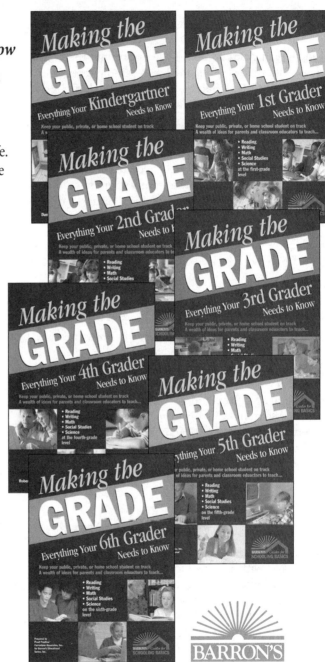

Barron's prepares students for the New Jersey ASK Tests!

Barron's New Jersey ASK3 Math Test

Thomas P. Walsh and Dan M. Nale

This brand-new book prepares New Jersey third graders for the ASK math test by explaining to teachers and students the test's format and all topics covered. It also points out the test's standards. Detailed review is given to the test's two question types, multiple-choice and open-ended, with advice on how to approach them. Two full-length tests are also presented with answers. This book's authors are an experienced elementary math teacher and a college math education professor.

Paperback, ISBN-13: 978-0-7641-3923-9

Barron's New Jersey ASK4 Language Arts Literacy Test

Lauren Filipek

This brand-new manual prepares fourth grade students in New Jersey to pass the required Assessment of Skills and Knowledge with a test explanation and overview, four chapters of instruction and exercises, and two full-length practice tests with answers. Fourth grade students are instructed in writing about pictures, reading and responding to a narrative (*i.e.*, story) text, writing from a poem prompt, and reading and responding to an everyday (*i.e.*, nonfiction) text. Study units are complemented with multiple-choice and open-ended questions for practice and review.

Paperback, ISBN-13: 978-0-7641-3789-1

Barron's New Jersey ASK6 Math Test

Mary Serpico, M.A.

This manual reviews the sixth grade math curriculum for the required New Jersey State math assessment test (ASK—Assessment of Skills and Knowledge). Five subject review chapters cover the following topics: number and numerical operations; geometry and measurement; patterns and algebra; data analysis, probability, and discrete mathematics; and integrating the strands through mathematical processes. Two full-length practice tests are included with answers.

Paperback, ISBN-13: 978-0-7641-3922-2

Barron's New Jersey ASK7 Math Test

John Neral

Seventh graders preparing to take the New Jersey ASK math test will find a detailed review of all relevant math topics, in this brand-new manual. Topics include integers, fractions, decimals, whole numbers and exponents, ratio and proportion, solving algebraic equations, patterns in algebra, pattern problems, inequalities, and much more. Each chapter includes practice and review questions with answers. Also included are two full-length practice tests with answers.

Paperback, ISBN-13: 978-0-7641-3943-7

Barron's Educational
Series, Inc.
250 Wireless Blvd.
Hauppauge, NY 11788
Order toll-free: 1-800-645-3476
Order by fax: 1-631-434-3217

In Canada:
Georgetown Book Warehouse
4 Armstrong Ave.
Georgetown, Ont. L7G 4R9
Canadian orders: 1-800-247-7160
Fax in Canada: 1-800-887-1594

(#156) R 1/08

Please visit www.barronseduc.com
to view current prices and to
order books